YALE AGRARIAN STUDIES SERIES

James C. Scott, series editor

SMART ALLIANCE

How a Global Corporation
and Environmental Activists
Transformed a Tarnished Brand

J. GARY TAYLOR
and PATRICIA J. SCHARLIN

Yale University Press / New Haven and London

Designed by Mary Valencia
Set in Veljovic type by Keystone Typesetting, Inc.
Printed in the United States of America.

Library of Congress Cataloging-in-Publication Data
Taylor, J. Gary.
Smart alliance : how a global corporation and environmental activists
transformed a tarnished brand / J. Gary Taylor and Patricia J. Scharlin.
p. cm.
Includes bibliographical references (p.   ) and index.
ISBN 0-300-10233-X (alk. paper)
1. Chiquita Brands International. 2. Rainforest Alliance. 3. Banana trade—
Environmental aspects. I. Scharlin, Patricia J. II. Title.
HD9259.B3C578 2004
338.8'8734772—dc22
2003061370

A catalogue record for this book is available from the British Library.

The paper in this book meets the guidelines for permanence and durability of
the Committee on Production Guidelines for Book Longevity
of the Council on Library Resources.

10   9   8   7   6   5   4   3   2   1

To our children—Geoffrey, Adam, and Joshua Taylor;
Wendy Moore, Janet Solomon, and Peggy Rambach.
We hope that, at last, this book will give them
some idea of what we do all day. May each of them find
a path in life that is as rewarding as ours has been.
We also dedicate this volume in special memory of the late, gifted
Luke Taylor, a sadly missed inspiration to our entire family.

Song of the Banana Man
by Evan Jones

Touris, white man, wipin his face,
Met me in Golden Grove market place.
He looked at m'ol' clothes brown wid stain,
An soaked right through wid de Portlan rain,
He cas his eye, turn up his nose,
He says, "You're a beggar man, I suppose?"
He says, "Boy, get some occupation,
Be of some value to your nation."
I said, "By God and dis big right han
You mus recognize a banana man . . .

"I leave m'yard early-mornin time
An set m'foot to de mountain climb,
I ben m'back to de hot-sun toil,
An m'cutlass rings on de stony soil,
Ploughin an weedin, diggin an plantin
Till Massa Sun drop back o John Crow mountain,
Den home again in cool evenin time,
Perhaps whistling dis likkle rhyme,
(Sung) Praise God an m'big right han
I will live an die a banana man . . .

"De bay is calm, an de moon is bright
De hills look black for de sky is light,

Down at de dock is an English ship,
Restin after her ocean trip,
While on de pier is a monstrous hustle,
Tallymen, carriers, all in a bustle,
Wid stems on deir heads in a long black snake
Some singin de songs dat banana men make,
Like, (Sung) *Praise God an m'big right han
I will live an die a banana man.*

"Den de payment comes, an we have some fun,
Me, Zekiel, Breda and Duppy Son.
Down at de bar near United Wharf
We knock back a white rum, bus a laugh,
Fill de empty bag for further toil
Wid saltfish, breadfruit, coconut oil.
Den head back home to m'yard to sleep,
A proper sleep dat is long an deep.
Yes, by God, an m'big right han
I will live an die a banana man.

"So when you see dese ol clothes brown wid stain,
An soaked right through wid de Portlan rain,
Don't cas your eye nor turn your nose,
Don't judge a man by his patchy clothes,
I'm a strong man, a proud man, an I'm free,
Free as dese mountains, free as dis sea,
I know myself, an I know my ways,
An will sing wid pride to de end o my days
(Sung) *Praise God an m'big right han
I will live an die a banana man.*"

Contents

Contents

Appendixes

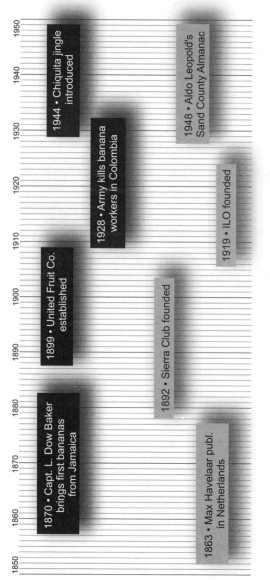

1850   1860   1870   1880   1890   1900   1910   1920   1930   1940   1950

1863 • Max Havelaar publ. in Netherlands

1870 • Capt. L. Dow Baker brings first bananas from Jamaica

1892 • Sierra Club founded

1899 • United Fruit Co. established

1919 • ILO founded

1928 • Army kills banana workers in Colombia

1944 • Chiquita jingle introduced

1948 • Aldo Leopold's Sand County Almanac

1848 • Communist Manifesto
1856 • London water pollution law passed
1863 • Emancipation Proclamation
1864 • Geo. Perkins Marsh (Man & Nature)
1866 • Telegraph connects US & Europe
1869 • Suez Canal
1872 • Minor Keith plants bananas in Central America
1886 • AFL formed
1890 • US Antitrust Act

1894 • National Audubon Society founded
1895 • Zemurray enters banana business
1900 • ILGWU formed
1901 • Marconi wireless across Atlantic
1902 • Alliance of Women Suffrage
1903 • US sovereignty over Panama Canal
1908 • Swedish chemist warns of greenhouse effect from fossil fuels
1911 • Triangle Shirtwaist fire

1920 • Ghandhi begins civil disobedience
1929 • Great Depression begins
1930 • Geest founded
1935 • Social Security Act
1938 • Fair Labor Standards Act
1945 • World War II ends
1946 • Largest strike wave in US history
1947 • Guatemala Labor Code passed
1949 • ICFTU founded

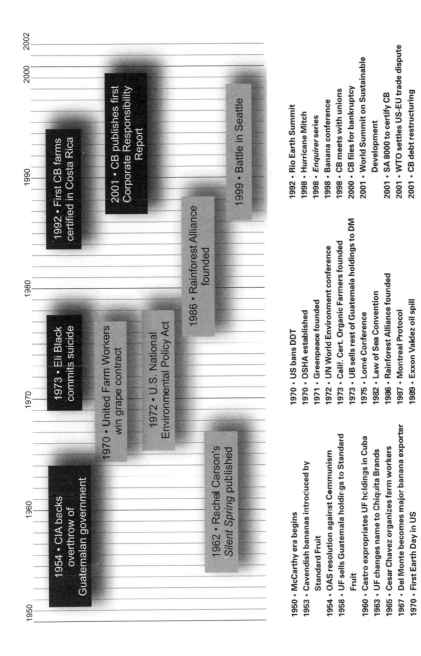

1950 • 1960 • 1970 • 1980 • 1990 • 2000 • 2002

1954 • CIA backs overthrow of Guatemalan government

1962 • Rachel Carson's *Silent Spring* published

1970 • United Farm Workers win grape contract

1973 • Eli Black commits suicide

1972 • U.S. National Environmental Policy Act

1986 • Rainforest Alliance founded

1992 • First CB farms certified in Costa Rica

2001 • CB publishes first Corporate Responsibility Report

1999 • Battle in Seattle

1950 • McCarthy era begins
1953 • Cavendish bananas introduced by Standard Fruit
1954 • OAS resolution against Communism
1958 • UF sells Guatemala holdings to Standard Fruit
1960 • Castro expropriates UF holdings in Cuba
1963 • UF changes name to Chiquita Brands
1965 • Cesar Chavez organizes farm workers
1967 • Del Monte becomes major banana exporter
1970 • First Earth Day in US

1970 • US bans DDT
1970 • OSHA established
1971 • Greenpeace founded
1972 • UN World Environment conference
1973 • Calif. Cert. Organic Farmers founded
1973 • UB sells rest of Guatemala holdings to DM
1975 • Lomé Conference
1982 • Law of Sea Convention
1986 • Rainforest Alliance founded
1987 • Montreal Protocol
1989 • Exxon Valdez oil spill

1992 • Rio Earth Summit
1998 • Hurricane Mitch
1998 • *Enquirer* series
1998 • Banana conference
1998 • CB meets with unions
2000 • CB files for bankruptcy
2001 • World Summit on Sustainable Development
2001 • SA 8000 to certify CB
2001 • WTO settles US-EU trade dispute
2001 • CB debt restructuring

Preface

This book grew out of our long-time efforts to understand the bound-aries between multinational companies and other central insti-tutions of our time. Our focus has been the governmental and non-governmental organizations concerned with global ecosystem integrity, sustainable development, and human rights. At the Sierra Club interna-tional office, we worked together in the eighties on campaigns to battle tropical deforestation, prevent toxic chemical proliferation in develop-ing countries, and find common ground between multinational com-panies and activists. With Tufts University we convened a series of work-shops at the Tufts international studies center in Talloires, France. These meetings brought together senior representatives from the public and private sectors to explore ways to jointly address important global en-vironmental and social issues. We found that spending several days shar-ing meals and ideas softened the edges of knee-jerk antagonism.

As the consulting principal investigator in a three-year research proj-ect for the Center for Environmental Management at Tufts, Gary Taylor examined the practices of a number of multinational companies at their headquarters and at facilities in this country and overseas. For ten years we published a commercial newsletter for executives in multinational companies and research organizations on practical issues of corporate environmental management. We interviewed senior company man-agers, officials of nongovernmental organizations (NGOs), regulators, and others in a biweekly tracking of multinational companies as they groped toward sustainable objectives. Too often, when companies began to publish stand-alone environmental reports to supplement annual re-ports, we found the companies declaring their good intentions by placing beautiful photos of birds or wilderness on the covers of their annual reports without providing honest and accurate data inside. We often joked that the more birds there were on the cover, the flimsier the data were likely to be. With our experience as environmental advocates and researchers, we were especially sensitive to "greenwash."

Eventually we came to believe companies and NGOs could work together in addressing environmental and social issues. We also became personally acquainted with people both inside and outside many companies. The work was gratifying, especially as we confirmed how similar human beings are, no matter where they labor.

We concluded that there was a paucity of credible, independent investigations of actual company operations—the "ground truth" of corporate practice on these issues. We were presented with an unusual opportunity to pursue this important subject when we learned that the Rainforest Alliance in New York, a group we knew well, had been working on the farms of Chiquita Brands International in Latin America and elsewhere, to certify that these operations were being run in an environmentally sound way—and that health and safety measures for the workers were in place. As a result of this novel collaboration (and other crucial events deeply affecting the company), this formerly secretive corporation was willing to open its facilities to us as well. It was a challenge we believed we had the experience and skill to take on, and we thought that, published as a book, our investigations would be useful—especially since the surprising relationship between Chiquita and the Rainforest Alliance was taking place at a time when multinational companies, along with many other social institutions, seemed to be losing their way.

The breakdown in the public's trust in corporations, governments, and religious organizations has left painful disconnects between society and these fundamental institutions. The road toward corporate responsibility for global companies has not been easy; and the critics have been many and vocal. The players involved in our drama—from Catholic priests in Central America to workers' rights campaigners and supermarkets in Europe—represent forces for change that are on the front lines of some of the major global economic battles of our time. The burden of proving that change is possible in a company such as Chiquita fell on the shoulders of a small group of people inside the company, and on the Rainforest Alliance. This is basically their story.

We started our investigations in Central America and focused on Costa Rica, for a number of reasons. It was there that the Rainforest Alliance's certification principles were beginning to take shape in the early nineties, and it is the headquarters of both the Rainforest Alliance and Chiquita's Latin American operations. It is also where a tremendous amount of relevant research was going on, and the government was open and cooperative. We describe the unfair conditions of the workers and the degradation of tropical forests by the venturesome banana entrepreneurs from North America and Europe who dominated the Central American isthmus at the start of the twentieth century. Interviews there at the end of the century helped us to explain how embedded the banana story is in the bigger dilemma of globalization and free trade. The dominant free-trade model had done much to marginalize the small farmer and to highlight the urgency of reform in the banana industry. Chiquita's decision to travel alone down the new path of environmental and social responsibility thus becomes a matter of international significance.

In Europe, we interviewed company, activist, research, and government people, all of whom play a major role in this story. In fact, understanding European interests is central to comprehending the forces that continue to affect corporate strategic thinking under way. So far, at least in the United States, the untouched ingredient in unleashing consumer power is the strong buying capability of the huge supermarket chains—companies much larger than Chiquita, such as Wal-Mart and Royal Ahold. Conglomerates continue to buy mostly on price. They have yet to demonstrate the kind of corporate responsibility that would unblock the bottleneck that is in danger of frustrating all the pioneering work we report. Fortunately, social investment groups are already laying the groundwork to call these companies to account.

This book, unlike many case studies of corporate behavior, is written not only for business managers and environmental and social professionals. We hope also to inform and encourage consumers everywhere. As members of the global community, they are the last vital links in the food chain and potential revolutionaries for the Earth and its people.

Many people helped us to research and write this book. Without doubt the most important are the small group of committed and hardworking people in the Rainforest Alliance and Chiquita Brands International, and the relatively small circle of activists, union leaders, and others in Europe, Latin America, and the United States whose interviews are at the heart of the drama we describe. These actors speak for themselves in our book, but we wish to single out Chris Wille and Dave McLaughlin, who gave us a tremendous amount of their time, thought, and feedback. They are truly heroes.

Many others aided us behind the scenes. Mark Halle encouraged our earliest efforts. Hector Brenes helped organize our very tight schedules in Central America. David Dudenhoefer provided excellent interpretation and a journalist's insight into the history and politics of that part of the world.

At Chiquita, we are grateful for the remarkably free access we were given. The company's policy of candor and forthrightness is the reason this story is worth telling. We particularly thank Magnes Welsh for her help in setting up our interviews with busy executives and for sharing her extensive knowledge of company operations.

Others who played a role in bringing this book to fruition include Gerardo Budowski, director of natural resources at the University for Peace; Bessie Vaquerano Castro of the toxic chemicals project in Costa Rica for the World Health Organization; Carlos M. Chacón from CEDARENA; Stephen Coats of U.S./LEAP; Liem Giok at the Management Projects for Individual Empowerment and Democratic Development; James Howard at the ICFTU; Bertius J. Meins, representative of the Inter-American Development Bank; J. Francisco Oreamuno, director of Costa Plant B.V. in Rotterdam; Nelcia Robinson from the Caribbean Association for Feminist Research and Action; Carlos Manuel Rodríguez, then the vice minister of the Ministry of Environment and Energy; José María Rodríguez from the Organization for Tropical Studies; Lorena San Román at the Earth Council; Lucien Royer, health, safety and environmental officer, ICFTU, and member of the trade union advisory committee to the Organization for Economic Cooperation and Development; Clemens Ruepert, environmental chemist at National University in Costa Rica; Ricardo Russo, professor of area natural resources at the research college

EARTH; Albert Schram, environmental economist; Lori M. Wallach from Public Citizen; and Alberto Salas and Florangel Villegas at the Meso-american regional office of the World Conservation Union (IUCN). The late John Wirth, professor of history at Stanford University, provided a forum for examining our findings with members of the North American Society of Environmental Historians.

In the initial days of the Ground Truth project that gave birth to this book, we had support from Jeremiah B. McKenna and Frank Penna of the Policy Sciences Center at Yale University. Early planning grants for the research came from Procter & Gamble, the Bristol-Myers Squibb Foundation, and the Canadian Pulp and Paper Association. For this support we are grateful to George Carpenter, Christopher Davis, and Thomas Hellman, all of whom saw the importance of telling the story.

Along the way we had the advice and counsel of a distinguished group of advisors. We thank them again for their early support of Ground Truth and the insights they brought from their sectional perspectives: environmental and social activism, corporate environmental management, and intergovernmental policy. We are grateful to the Conference Board in New York for providing meeting space.

Numerous friends and colleagues listened patiently for more than three years as we talked incessantly about bananas; they supplied additional information, insight, and expertise. Among them are Mark Nedostup, who calmly pulled together all our illustrative material, and Lawrence S. Hamilton, who called our attention to early forest management efforts. Noel J. Brown's sophistication and knowledge of the United Nations was essential, as were the wise and informed business perspectives of David Lowy, Fritz Gold, and Bill Stibravy. Sister Mary Theresa Plante from Franciscans International was a reliable source on labor issues.

Eben Forbes and Shawnee Hoover, Columbia University graduate student interns, played helpful roles in researching details and organizing our information. We deeply appreciate their dedication to the field they have entered and we wish them great success and professional gratification.

Our agent, Joe Spieler, kept the faith in our darkest moments. The enthusiasm and graceful editing of our editors at Yale University Press,

Jean Thomson Black and Vivian Wheeler, gave us courage, too. Margaret Otzel found a way to organize our final changes in this ongoing story.

Unless credited to other sources, the photographs and other illustrative materials were contributed by Chris Wille, the Rainforest Alliance, Chiquita Brands International, and the authors. Naturally, we authors are responsible for any factual errors or omissions.

Finally, full disclosure demands that Pat's years on the Board of the Rainforest Alliance be noted. While this sharpened the need for objectivity, it also enabled us to provide an insider's perspective. The book has benefitted enormously from the unusual candor and trust in the culture of this adventurous group.

Abbreviations

ACCA	Association of Chartered Certified Accountants
ACP	Former colonies of countries of the European Union in Africa, the Caribbean, and the Pacific receiving special treatment under the 1975 Lomé Convention
ATO	Alternative Trade Organizations
BBP	Better Banana Program of the Rainforest Alliance
BSR	Business for Social Responsibility
CCT	Centro Cientifico Tropical, a Costa Rican research organization
CEC	Commission on Environmental Cooperation (a body created under a NAFTA side agreement)
CEP	Council on Economic Priorities
CERES	Coalition for Environmentally Responsible Economies
CI	Conservation International
Codex Alimentarius	a UN commission of the Food and Agriculture Organization and the World Health Organization established to harmonize standards on definitions and requirements for foods in order to facilitate international trade
COLSIBA	Coordinadora Latinoamericana de Sindicatos Bananeros, the coordinating organization of Latin American banana workers' unions representing some forty trade unions in eight countries

CORBANA	National Banana Corporation, Costa Rica
EARTH	La Escuela de Agricultura de la Región Tropical Húmeda (Agricultural College of the Tropical Humid Region)
EEC	European Economic Commission
EMAS	Eco-Management and Auditing Scheme
EPA	Environmental Protection Agency
ETI	Ethical Trading Initiative
EUREP-GAP	a recently formed European supermarket certification program
EUROBAN	the European Banana Action Network, a confederation of some thirty European nongovernmental organizations and trade unions from twelve countries that work together on workers' and environmental rights and promote Fair Trade bananas
FDB	a Danish supermarket cooperative chain
FLO	Fair Trade Labeling Organization International
GATT	the General Agreement on Tariffs and Trade
GEMI	Global Environmental Management Initiative
GM	Genetically modified
GRI	Global Reporting Initiative
IBIS	U-landsorganisation Ibis (a Danish environmental education nongovernmental organization)
ICCR	Interfaith Center for Corporate Responsibility
ICFTU	International Confederation of Free Trade Unions
IFOAM	International Federation of Organic Agriculture Movements
ILO	International Labour Organization
INBio	National Biodiversity Institute, Costa Rica
iSEAL	International Social and Environmental Accreditation and Labeling Alliance
ISO	International Organization for Standardization
IUCN	World Conservation Union (originally the International Union for Conservation of Nature and Natural Resources)

IUF	International Union of Food, Agriculture, Hotel, Restaurant, Catering, Tobacco, and Allied Workers' Associations
NAFTA	North American Free Trade Agreement
SAI	Social Accountability International
SAN	Sustainable Agriculture Network
SASA	Social Accountability for Sustainable Agriculture
SGS group	a private-sector global standards and certifications organization
SID	Danish General Workers' Union (SiD)
SITRAP	Sindicato de Trabajadores de Plantaciones Agricolas, the banana workers' union in Costa Rica headed by Gilberth Bermudez
UNCTAD	United Nations Conference on Trade and Development
UNEP	United Nations Environment Programme
USAID	U.S. Agency for International Development
U.S./LEAP	U.S. Labor Education in the Americas Project, a nonprofit organization for education and advocacy on Central American labor issues
WIBDECO	Windward Islands Banana Growers' Association
WSSD	World Summit on Sustainable Development
WTO	World Trade Organization

Introduction

There are seven sins in the world:
Wealth without work,
Pleasure without conscience,
Knowledge without character,
Commerce without morality,
Science without humanity
Worship without sacrifice, and
Politics without principle.

Mahatma Gandhi

S tories about responsible companies are rare these days. Highly visible companies like Enron have betrayed investors and workers, giving the concept of certifiable corporate probity a new urgency that goes beyond the halting reform of financial accounting.

Many of the largest and wealthiest multinational companies have failed to acknowledge their responsibility to address global environmental degradation and workers' rights issues. But formal acknowledgment of corporate responsibility is only a first step. What is called for is a tough and credible means of assessing corporate performance "inside the factory gate" in order to verify management's commitment.

When so many big companies have been heading downhill in these respects, astonishingly, Chiquita Brands International—successor to the much-reviled United Fruit—has been heading upward in a steep trajectory that almost defies credibility.

Bananas have their comical side, but the export banana game is a huge and very serious global business. It encompasses a vast area of land in the tropics; hundreds of thousands of workers; millions of boxes of bananas shipped to Europe, Japan, and North America; use of untold gallons of freshwater and tons of agricultural chemicals. In terms of volume, the droll, sexy, and nutritious banana is the most exported fruit in the world.

This book tells how building trust and taking risks—the essential ingredients of reform—evolved from the unexpected alliance between Chiquita and an upstart environmental organization, the Rainforest Alliance. Their core idea was an innovative environmental "seal of approval." The acknowledged improvements that resulted from the unlikely collaboration leveraged a series of internal and external changes in Chiquita's financial and social commitments and triggered a transformation in the entire banana industry. If demonstrable and lasting improvements ultimately result, the innovation could become a model for other industries and nongovernmental organizations willing to take chances in order to overcome historic barriers of mutual distrust.

Though voluntary partnerships will never replace the rule of law under fairly elected governments, our account shows how a customer-driven tool, in this case certification, can create change in complacent corporate cultures. This type of motivation can be especially effective in countries with limited law enforcement resources and questionable will to stand up against the multinational companies on which their economies have become so dependent.

Our year-by-year inside account tells of a crucial turning point in the history of a company originally built on sheer size and arrogance. With the goading and help of the Rainforest Alliance, Chiquita took a number of dramatic steps to shed its unsavory past and put itself squarely on the road to being that rarest of corporate breeds: a bona fide responsible corporation—environmentally, socially, and financially.

Two emerging trends come together in this tale. The first is the radical idea of allowing an independent party to certify the environmental and social performance of multinational companies. The second is the slow but promising corporate movement toward voluntary, candid corporate reporting. Together these trends have at least a chance of breaking the logjam created when world trade and globalization confront environmental and social imperatives.

When it is truly independent and credible, certification can unleash the power of consumers, enabling them to make informed choices among goods produced continents away. Increasingly, informed shoppers do not want to buy products that they know have been produced in sweatshops, or from farms that can succeed in the global marketplace only by cutting corners on labor rights and environmental safeguards. When they are reliably informed, consumers can be transformed from passive victims of corporate spin into a legitimate political force focused on making responsible companies serious agents for positive societal change.

A new factor in this revolution is corporate dedication to burnishing global brands that are now supervulnerable to "Internetworked" protests around the world. Any company's coveted brand can be trashed on computers in every corner of the globe. The risk of having one's company ridiculed on computer screens in a matter of minutes is an issue-management nightmare. Nongovernmental organizations too have become sensitized to the double-edged sword of branding. Vociferous protests in Seattle in 1999 put these issues on the television screen, and the protests at all international meetings on trade, economic policies, human rights, and the environment are reaching the public as never before. Activists of all kinds have become more sophisticated in presenting their views, and the behavior of a growing number is more responsible than ever before. The best of these activists are increasingly being brought to the tables where public policy is forged. And the multinational companies that have learned how to work with these groups are finding ways to be profitable and responsible at the same time.

It is a two-way street, however. Activists—including certifiers—have responsibility to maintain a clear, transparent, and independent identity. Consumer choices, especially in rich countries, can be overwhelming. Labels that certify products need to explain clearly to the buyer what they stand for. Otherwise they will be replaced, inevitably, by government standards vulnerable to lobbying pressures, and goods will be produced with little regard for people or the planet.

During the 1980s public outrage ignited over tropical forest destruction, worker illness from overuse of pesticides, and visible damage to coastal ecosystems. Almost all the big agricultural companies organized campaigns of denial. Chiquita broke ranks and allowed the dedicated environmentalists at the Rainforest Alliance onto its farms in the tropics to certify that the company's performance was improving every year. Managers were measurably reducing the use of toxic chemicals, replanting the rainforest, improving the equality of ground and surface water, and making costly changes to protect workers' health and enhance their safety. In this highly competitive industry, the relationship astounded Chiquita's U.S. rivals, Dole and Del Monte, and worried conservationists and other critics who needed proof that the Rainforest Alliance had not sold out.

Chiquita and the Alliance worked without public fanfare since 1992 to improve environment, health, and safety performance on all of the farms owned by the company in Central and South America. In 1998, driven by a variety of internal and external forces, Chiquita resolutely took more public steps. It initiated a series of escalating commitments to integrate an ambitious program of corporate responsibility into its overall management structure, adopting what it described as a set of "core values."

At the center of the new program was a tough labor standard to guarantee fair and equitable working conditions on all Chiquita farms, and the decision to use an independent former corporate watchdog organization to publicly benchmark social commitments based in part on conventions established by the International Labour Organization. In an even more startling move, the company in 2001 signed a pioneering agreement with an international umbrella organization of workers in the food industry and also with a Latin American labor coalition of banana workers. The agreement bound the company to a series of steps that won praise even from some of Chiquita's harshest critics. Remarkably, Chiquita sustained its commitment through a debilitating bankruptcy, a complete change of ownership, and a new chief executive officer.

In this volume we focus on the early years of Chiquita's difficult work with the Rainforest Alliance, since those efforts became the linchpin of the company's subsequent breakthrough in labor relations. Chiquita employs some twenty thousand workers, the largest number of agricultural laborers in Latin America. Over its hundred-year history, it had acquired a notorious record of unsavory habits. When a single company can give an entire region its reputation as a collection of "banana republics," it has much to live down.

On the environmental front, improvements have been slow but meaningful on all of Chiquita-owned farms. Moreover, the company and the Alliance are well on their way to fully certifying independent farms that sell bananas to Chiquita—a daunting challenge indeed. We personally testify to improvements in worker health and safety measures on these farms as well. Although most of Chiquita's commitments to workers' rights must still be demonstrated, the company has already established a profound precedent by documenting in fine-grained public detail the benchmarks and the distance to go to achieve the targets. As it turns out, the Rainforest Alliance, which went into the process thinking of itself as a catalyst, has emerged a changed organization. Undeniably it has learned and grown along with Chiquita.

One lesson we hope readers take from this book is that even if they cannot tell one banana from another, now they can easily tell one banana brand from another. The story behind the label is what counts: it has to do with risk, transparency, and trust.

1

Risk, Transparency, and Trust

In early March 1999—exactly one hundred years after the birth of United Fruit—we were in Costa Rica on our way to banana country. We were accompanied by David McLaughlin, then senior director environmental affairs, tropical environmental group, for Chiquita Brands International, the successor company to United Fruit. This company, above all others, had dominated the banana business in Latin America. Even if the company had been in sync with public expectations in its early years, by the closing decades of the twentieth century it had lagged far behind rapidly changing global norms. We were after "ground truth" on environmental and social changes in the company and the role that the Rainforest Alliance, a young conservation group, was playing in these changes that some were already describing as a transformation. We wanted to talk to critics, government officials, scientists, labor leaders, workers, and anyone else who could shed light on whether the notorious banana company, dubbed El Pulpo (the octopus), was truly changing its ways. We also were eager to see for ourselves the improvements that we had heard so much about.

We were soon to meet the people who jump-started this process of change: conservationist; a seasoned agriculture expert; a young former scholar of Chinese; a businessman and orchid fancier; a few practical but visionary executives at Chiquita; a cool, smart linguist and small-farmer advocate; dogged workers' rights activist groups, including one in the American Midwest; a few trade union leaders in Europe and Central America; and religious and social investment watchdogs.

What were their instruments of reform? Old-fashioned campaign tactics, coupled with a new tool: independent assessment on the ground of the production of bananas, to ensure equitable treatment of workers and protection of the land and surrounding environment.

Why did their approach work? Because in our brand-driven global economy, banana customers and activists in northern Europe cared about how bananas are grown and how workers are treated thousands of miles away in the tropics. They rallied behind a protectionist European Union quota, tariff, and license system that sharply cut Chiquita's potential banana market in the newly unified Europe in favor of exports from existing territories and former colonies. Chiquita's business rivals in Europe completed the composition of forces.

What was the unique blend of ingredients and unexpected synergies that wrought such a powerful company change? A new generation of managers at Chiquita determined to shake off one hundred years of banana republic reputation in order to protect a valuable brand name; plus a young, inventive group of conservation risk-takers who realized they had an idea "so practical it's radical," as they said; plus, very recently, an independent social certification group. Together, they understood that business-as-usual was no longer working. They saw that global institutions, including national governments, could not stem the tide of environmental degradation or improve the lot of working men and women without major help from the private sector.

Of course, nobody thought then, or believes now, that independent environmental and social certification of banana farms—or for that matter, coffee, flowers, cacao, or oranges—is a panacea. And nobody believes that the huge infrastructure and behavioral changes required will happen overnight, or even over the course of many years. But the idea may have come along just in the nick of time. We were investigating the banana industry at a time when our Earth and its people were on a dangerous path of ecological disequilibrium, social volatility, and unrelenting poverty. We wanted to determine if certification, transparently and rigorously applied, could be an important new path, with an opportunity to make multinational corporations partners on the long path to sustainable development and to put consumers back in the driver's seat.

We must be careful not to overstate global environmental and social troubles, yet a report prepared by the United Nations Environment Programme (UNEP) for the World Summit on Sustainable Development, convened in Johannesburg, South Africa, in the late summer of 2002, spells out a number of very worrisome realities.[1]

Tropical forests are being deforested at a rate of almost 1 percent a year. According to the Rainforest Alliance, this is the equivalent of two U.S. football fields every second. Why is this depletion important? For many reasons. Some people want to protect the rainforest simply because it is there, a gift from God that it is our duty to preserve. Others who experience the sheer exuberance of nature in the tropical forest are concerned about the unthinking obliteration of plant and animal species, all of which play essential roles in the underlying ecological balance of life on Earth.

An unknown number of these species have already contributed dramatically to human health and potentially could contribute much more. From the Peruvian rainforest, the quinine plant is used in the treatment of malaria. From the tropical forest of Madagascar, the rosy periwinkle has proved to be a powerful anticancer agent. Almost fifty major drugs have been derived from rainforest plant species, at an estimated net worth of $147 billion.[2]

According to the 2002 UNEP report, global biological diversity is being lost at a rate many times higher than that of natural extinction, owing to land conversion, climate change, pollution, unsustainable harvesting of natural resources, and the introduction of nonnative species. About a thousand species of mammals and almost twelve hundred species of birds are thought to be headed for extinction barring some sort of intervention.

UNEP's report also tells us that about a third of the world's population lives in countries that have moderate-to-high water stress. By the mid-1990s, some eighty countries were suffering serious water shortages. Toxic pesticides and other hazardous chemicals kill or seriously sicken thousands of people every year. They also poison the natural environment and threaten many wild animal species.

Yet the picture is not one of total disaster. The report finds that in the last decade the European atmosphere improved significantly, and in fact overall environmental conditions are either stable or improved in both North America and Europe. Africa's wildlife conservation record is impressive. On the other hand, the "vulnerability gap" between well-off people with all-round coping skills and the many poor people who lack

Deforestation for banana farms in Costa Rica, 1992

many of these skills is widening. The report concludes: "Poverty and excessive consumption continue to put enormous pressure on the environment. The unfortunate result is that sustainable development remains largely theoretical for the majority of the world's population of more than 6,000 million people."

The social side of the equation is equally perilous. The 2001 annual report on violations of trade union rights published by the International Confederation of Free Trade Unions (ICFTU) painted a grim picture. Force was repeatedly used instead of dialogue. Employers or the police in nearly ninety countries around the world repressed more than three hundred strikes or demonstrations in which trade unions took part. In

Bolivia the police, during two major social movements calling for higher pay and lower water rates, killed at least fourteen people. In Costa Rica, Argentina, and Paraguay, demonstrators also lost their lives. In Bangladesh, four workers were killed by police in the port of Mongla while demonstrating to demand more staff.

Murder is only the most extreme reaction according to the report; employers often use hired thugs to mix with strikers so that the police will have an excuse to intervene. In Turkey in 2001, the government instituted legal proceedings or administrative inquiries against eighty-six thousand civil servants who took part in a strike it had declared illegal. In all the Latin American countries cited in the report, with the exception of Cuba, at least one strike was repressed in the same year. In Ecuador and Venezuela, the police intervened violently in no less than ten strikes. In May 2002, men claiming that the private Ecuadorian banana company Noboa had hired them attacked union workers shortly after the government had legally recognized their unions—a struggle that so far seems to have no easy resolution. In Europe, the right to strike is also problematic. In Belgium, employers frequently use the courts to put an end to strikes by imposing excessive restrictions on the pickets. While unions in the United States do not represent the majority of workers, public acceptance of the core labor standards set down in conventions of the International Labour Organization is growing. The right to freely organize, to collectively bargain, and the other rights and prohibitions enshrined in these conventions are still barely realized in many places in the world.

Tropical forest destruction, pesticide poisonings, and union repression have all been associated over a long past with the production of bananas and with all the major banana companies. Reforming this huge industry is no small undertaking, especially given the growing imbalance of resources between private sector and public sector, a dominant theme in the deep divide between those for and against globalization. A surprisingly potent symbol of the new set of issues being raised by global protesters has been the ubiquitous banana. And the brand associated with it in just about everyone's mind is Chiquita.

The sheer size and reach of multinational companies in the recent period are beyond question. A recent study by UNCTAD (the United Nations Conference on Trade and Development) found twenty-nine corporations that were worth more than many countries. The hundred largest corporations accounted for 4.3 percent of global gross domestic product in 2000. ExxonMobil, which is worth $63 billion, is bigger than Pakistan. General Motors, at $56 billion, outranks both Peru and New Zealand. Over the recent past, multinational company strategies to take advantage of broadened market access have generated new approaches to integrated manufacturing networks and marketing tie-ins that put a premium on global branding. The images are everywhere and make the reach of these companies, if not their actual presence, a global phenomenon impossible to ignore.

The globalization debate has been put dramatically: Is our wildly disparate world community being forged by ExxonMobil, General Motors, and their corporate brethren into a buoyant new trading system that, by mistake or design, will at long last erase the chasm between rich and poor countries? Or will these juggernauts of capitalism and greed, driven by quarterly earnings reports, grind the poor to dust under heedless, amoral heels? From this point of view, reform of multinational corporations is so unlikely that it is virtually counterintuitive.

Practically everyone agrees that growth is essential. But the global international institutional capacity to provide *equitable* growth is problematic. Most of the recipes to make globalization work for everyone involve reform of national governments, yet right now it is difficult to think of a Third World government, especially in the poorest countries, where that is an option. Consider Haiti. Some poor countries have improved their economies by importing assembly lines (in Mexico called *maquiladoras*). But these are likely to be a false hope unless the rich partner also transfers the technology. South Korea may be an example of this. Of course, reform would come more quickly if one could excise powerful private companies that muscle their solutions into acceptance by overwhelmed national bureaucracies.

Strengthening national governments, setting up a system to enable true technology transfer between the North and the South, or finding a way to dampen the ardor of the special or corrupt interests involves slow change—crawling along as the environment degrades, workers suffer, and the poor get sick and die. The main beneficiaries of globalization are still too often the relatively small number of elites in one or another developing country.

Growth in the Number of Multinational Corporations, 1600–2000

Number of Corporations:

Year: 1600   1700   1800   1900   2000

SOURCE: Medford Gabel and Henry Bruner, An Atlas of the Multinational Corporation *(New York: New Press, 2002). Courtesy of Global Inc. Also see www.bigpicturesmallworld.com.*

Still, corporations are not in lockstep. For a variety of reasons, including the desire to do "the right thing," a number of big companies are getting certified to an environmental business-to-business standard (known as ISO 14001), which involves an internationally accredited independent sign-off on a company's management system, though not on its actual performance. Other companies are adopting codes of conduct—some of them with real credibility. A growing number have joined a coalition of nonprofits to develop and adhere to a standard for voluntary annual environmental and social reports, a definite step in the right direction. Big companies are banding together in associations such as the Global Environmental Management Initiative (GEMI), the World Business Council for Sustainable Development, and the Prince of Wales Business Leaders Forum, seeking to find common ground between profit and social and environmental responsibility.

At this point in history, an elaborate feeling-out process is going on among governments, corporations, and nongovernmental organizations. Some enterprising NGOs are entering into limited partnerships with companies. These range from one-shot, project-oriented actions to longer-term, issue-based initiatives. Clearly, big corporations have the money and the power to be major change agents, especially as their dominance and influence begin to exceed those of many other institutions in the world. The trend for economically strapped cities to sell "naming rights" of companies such as Coca-Cola, Ford Motors, and Saturn to public parks, interpretive centers, and playgrounds is but one example of the ubiquity of corporate branding.

We are barely into the new millennium, and the picture has changed a great deal from the early days of the Rainforest Alliance's certification initiative with Chiquita. At first, by working with the markets instead of against them, the Alliance found itself needing to engage many of Chiquita's antagonists, especially European business interests and labor and environmental activists on both sides of the Atlantic. The two organizations had to deal with a European "not-invented-here" mindset and the all-too-familiar anti-American rhetoric. Today, pioneering American environmental and social certifiers are finding that they must come to terms with competing certifications like Fair Trade and organic labelers that have the advantage of a long religious and social reform history, especially in Europe.

Despite their critics and rivals, our protagonists have been able to weave nimbly around the sluggish governments and academic naysayers. There is something to be said for staying beneath the radar of the bureaucrats: the truth is that without the help of governments, aid agencies, or high-priced "experts," Chiquita and the Alliance have made a solid contribution. They have done so with amazing speed at a cost that would astonish the official development assistance operatives and some of the heavily funded international NGOs. Certification acting as a market incentive, as opposed to enforcement through governmental regulation, has also entered the political discussion in international forums dealing with the tropical timber trade, among others.[3]

The story of the Rainforest Alliance and Chiquita is an absorbing account in microcosm of the big forces driving the globalization debate. Both groups stayed the course despite a difficult trip through bankruptcy for Chiquita and a change of leadership at both organizations. As we write, all of Chiquita-owned farms have been certified, and 46 percent of all Chiquita's bananas from independent growers come from certified farms, up from 33 percent in 2002. Most of the animosity toward Chiquita from Europe and the NGO communities has become muted. In 2001 and 2002 the company's corporate responsibility reports included a groundbreaking new commitment to workers' rights. These take Chiquita light years beyond most other corporations and raise the bar substantially in its own industry.

2

Red, White, and Bruised

Respect voor mens en milieu

The 1990s were boom years for many multinational companies. They were also boom years for the restive protest movements around the world. But the decade was a roller-coaster ride for Chiquita. In the minds of the company's principal stockholder, Carl H. Lindner Jr., and his cadre of corporate strategists, these were to be the glory years. The vaunted European Common Market and the fall of the Iron Curtain would finally result in a huge share of the banana market in Europe, where the Chiquita brand could command a decisive premium. Chiquita would recoup major investments in new farms in Central America, refrigerator vessels, and ripening and other facilities, and go on to reap the major profits the company had been promising stockholders.[1]

Instead, in 1994, just when Lindner—the reclusive, devoutly Baptist multimillionaire and venerable Cincinnati benefactor—and his executive team were marshaling their forces for a major thrust, the European Union installed an unexpected and punishing trade system. Quotas and a system of licenses were designed to favor growers in former colonies and territories in the Caribbean and Africa at the expense of the big American dollar-banana companies: Chiquita, Dole, and Del Monte. The backbones of European trade officials were stiffened by activist groups, along with European business adversaries, who hammered Chiquita for its negative banana-republic reputation. During this period the company was quietly spending millions of dollars to secure green

Photo previous page: An example of the type of postcard European activists have used against dollar banana companies.

certification for its farms in the tropics through a barely known American environmental group, the Rainforest Alliance.

A Quadruple Whammy

In 1998 Chiquita was hit by four major events that were to mark the beginning of meaningful corporate reform. The first blow came on February 19, when nearly five thousand Panamanian workers went on strike against a Chiquita subsidiary, the Chiriquí Land Company. The workers demanded to use the Pacific coastal port of Puerto Armuelles instead of the current port on the Atlantic to ship the bananas, claiming that the change caused 275 banana workers to go unpaid for three months. The strike went on for fifty-seven days, despite mediation by Panama's labor minister. The workers' grievances included reduced per-piece salaries, because of the time it took to insert the special pads required to protect the bananas. Other grievances included the company's slow progress in improving housing and latrines. Ninety-eight percent of the banana workers supported the strike, which was declared legal by a Panamanian court on February 28 in response to a lawsuit by the company. The Chiriquí Land Company claimed that after the first ten days of the strike, some 360,000 boxes of bananas, worth $1.5 million, were not exported. Union leader Edgar Omar Williams said the proposals of Chiriquí Land Company were "vague and a bit disrespectful."

On May 3, as the company was coping with the aftermath of the strike, Cincinnati citizens opened their hometown newspaper, the *Cincinnati Enquirer,* to read a front-page headline: "Chiquita SECRETS Revealed; *Enquirer* investigation finds questionable business practices, dangerous use of pesticides, fear among plantation workers; Chiquita: An empire built on controversy." Inside was a series of articles in an eighteen-page special section written by reporters Mike Gallagher and Cameron McWhirter. Their information derived from a year's research in Central America, Europe, the Caribbean, and North America, scores of interviews, and excerpts from internal Chiquita voice-mail messages.[2]

The articles accused Chiquita of numerous nasty business practices: they charged that company executives had bribed government officials in Colombia; had poisoned plantation workers through careless use of pesticides, including aerial spraying; and secretly controlled scores of ostensibly independent companies to avoid taxes and subvert union contracts.

Fortunately for Chiquita, it turned out that Gallagher had gained access to the voice-mail tapes illegally, relying on cooperation from a company insider. Outraged, Chiquita moved in its heavy legal guns, but even before it brought any action, the *Enquirer* publicly apologized to Chiquita in front-page banner headlines for three days running, renouncing the series and admitting the tapes had been stolen. The paper also paid Chiquita a reported $14 million to avoid being sued, an amount that ranks among the highest ever paid by a news organization. Gallagher was fired and McWhirter was transferred to the *Detroit News*. The company at the time did not acknowledge the accusations, and knowledgeable critics have confirmed many of the charges. (The company, in its corporate responsibility report for 2001 [published in 2002], confirmed past brutality toward some workers.)

The renunciation and apology by the *Enquirer* have gone a long way toward discrediting the series, but there has never been a charge-by-charge company response. Moreover, many of the accusations against Chiquita could also have been made against other banana companies: practice in the tropics over the years has not varied substantially from one company to the next. The hard truth is that environmental and labor abuses on Chiquita farms had become the stuff of legend. If over the years workers had been maltreated or the environment dealt with recklessly, the odds are that Chiquita was at least as culpable as its rivals, even into the Lindner era. No matter that many of the charges referred to past practice; Chiquita's evasiveness and arrogance, even to the reporters, made the company the target of choice for a growing number of critics.

This barrage of adverse publicity was something most Chiquita executives understood to be much deeper than a public relations problem. For one thing, company lobbyists in Washington, pressing for action by the United States trade representative against the throttling European trade regime, would have major new allegations to explain away. Carl Lindner personally had over the years given substantial sums to both the Democratic and Republican parties. He could not afford to have his money seen as tainted or his friends in Congress compromised any more than they already had been by campaign-finance reform efforts.

In addition, charges against the Rainforest Alliance questioned the independence and rigor of the innovative third-party environmental and worker safety certification program that Chiquita had pioneered in 1992. The *Enquirer* broadside specifically noted that although the Rainforest Alliance certification banned all agricultural chemicals except those approved by U.S. and European governments, Gallagher and McWhirter had found that at least in one case a chemical was being applied that was not approved by one European country. A former staff member of the Alliance told us the episode taught him that the language on prohibited agrichemicals in the certification standards was too general. It called merely for the products to be registered for use on a particular crop and approved for this purpose by the Environmental Protection Agency in the United States and other agencies in Canada and Europe. It was a real "gotcha," and the Rainforest Alliance had to deal with it if it hoped for acceptance in the long run. The relationship that had been building between Chiquita and the Rainforest Alliance was undergoing its first test.

Despite the unusual apology and payment, the antibanana forces in Europe would use the allegations in the series to great advantage. In the third blow that month a major banana conference, convened in Brussels by the coordinating organization of Latin American banana workers' unions (COLSIBA) and a newly formed workers' rights umbrella group (EUROBAN), seized on the charges in the *Enquirer* as ammunition to bolster their increasingly rough campaign against Chiquita. Steven G. Warshaw, then president and chief executive officer of Chiquita Brands International, Inc., walked into the company's headquarters on East Fifth Street in downtown Cincinnati and announced the launch of a top-to-bottom program of corporate responsibility. Warshaw, perhaps cut from somewhat different cloth than earlier Chiquita bosses, would not allow the newspaper and the activists to have the last word on a company that, after all, was under management a few generations removed from the old United Fruit days. Warshaw firmly believed that the onerous European trade regime would eventually be ruled illegal. The World Trade Organization (successor to the General Agreement on Tariffs and Trade, or GATT) had already done so a number of times before the European Union finally backed down in 2000. He knew that Chiquita's economic future depended on a clean, "fun and healthy" brand—and on committed employees who needed to believe they were working for one of the finest companies in the world.

Warshaw had actually been forewarned that the *Enquirer* story was in the works: the reporters' questions had been funneled through the corporate law department and hence certainly to top management. Interestingly, Robert H. Dunn, chief executive officer of Business for Social Responsibility (BSR), told us that as early as April 1998 his nonprofit San Francisco-based business advisory organization had received feelers from people inside Chiquita. So even as the reporters were listening to company voice tapes, Warshaw had begun to consider radical new options.

Chiquita's 1998 woes would not end with the banana strike, the *Enquirer* fracas, and the anti-Chiquita campaign in Europe. Instead, starting around midday on October 27, the center of Hurricane Mitch meandered near the north coast of Honduras. It made landfall during the morning of October 29 about 70 miles east of La Ceiba, with estimated surface winds of 85 knots and a minimum central pressure of 987 millibars. The storm dealt damaging blows to Central America's banana-dependent countries. In Honduras alone, officials reported an estimated seven thousand people dead, one million driven from their homes, most highways and bridges damaged. The banana industry in Honduras, which represented nearly 20 percent of the world's supply, lost 70 percent of its crop. Chiquita, with 35 percent of its banana production in that country, announced that it was laying off all its 7,800 workers at banana plantations for at least a year. A Chiquita subsidiary, Tela Railroad Company, one of the nation's largest private employers, estimated that its plantations had suffered $200 million in damage and would be unable to reach prestorm production levels for at least three years. Mitch, deemed the worst disaster in a hundred years of Honduran history, also caused flooding, mudslides, loss of life, homelessness, and widespread crop damage in Nicaragua, El Salvador, and Guatemala. Chiquita joined a wide range of public and private organizations in the relief effort.

The *Enquirer* story and Mitch together produced immediate shock waves in Cincinnati that gave new urgency to the tentative corporate reform thinking going on inside the company. After the newspaper series broke, the Interfaith Center on Corporate Responsibility (ICCR), a coalition of mainly church-based groups founded to advise and coordinate share-holder campaigns against company social policies, contacted Sister Ruth Kuhn of the Sisters of Charity in Cincinnati. Sisters of Charity is itself a member of a coalition of religious institutions that invest in companies in order to exert pressure on them to reform. In fact, Sisters of Charity had owned Chiquita stock for some time before the fateful year 1998. At the center's request, other members of ICCR purchased Chiquita stock to support efforts to pressure the company. They wrote letters asking about company practice on their farms, especially in Honduras. Then came Hurricane Mitch. Sister Ruth says: "It was a turning point for us and for Chiquita. The company got some good publicity when Warshaw visited the farms, agreed to their rehabilitation, and kept many workers on the payroll. We wrote letters thanking the company for its efforts after Mitch and asking it to enter into negotiations with COLSIBA."

Sister Ruth addressed the Chiquita board of directors on these matters at the annual general meeting. At its conclusion Warshaw approached Sister Kuhn and spoke with her about the company's policies. She says this impressed her; CEOs almost never do that sort of thing. In fact, prior to Warshaw's overture, the Sisters of Charity had received no response to their letters. But shortly afterward, the company sent a three-page letter to the Sisters of Charity describing its new programs. Sister Ruth says, "There's no doubt in my mind the company did it purely as good busi-ness practice, not because it came from a moral stance." Whatever the motives, at that point a corporate responsibility function was established inside the company, under the up-and-coming strategic analyst Jeffrey M. Zalla. Sister Ruth says, "Mitch was truly devastating, but in view of the changes it triggered in Chiquita it was a wakeup call from the Lord."

The Sisters of Charity and the ICCR were not alone in the growing ant-Chiquita campaign. Stephen Coats and his colleagues at the U.S./ Labor Education in the Americas Project (U.S./LEAP) were working to support the basic rights of workers in Central America, Colombia, Ecua-dor, and Mexico, especially those who were employed directly or indi-rectly by U.S. companies. Founded in 1987 as the U.S./Guatemala Labor Education Project (U.S./GLEP) by trade unionists and human rights advocates concerned about the basic rights of Guatemalan workers,

U.S./LEAP had expanded its efforts to other countries in the region. Chiquita during this period was frequently featured in the organization's broadsides and on its Internet site.

Weeks before Mitch struck, Warshaw had initiated an intense exploratory session with Business for Social Responsibility. In August 1998, in the wake of the turbulent events earlier that year, he traveled to San Francisco to meet with the CEO of BSR, Robert Dunn. "What was extraordinary," Dunn recalled when we talked with him at his headquarters in late November 2002, "was that when we meet with a CEO in person, he usually wants us to visit him. In this case Warshaw came to us. After a very short exchange of pleasantries, Warshaw reached into his briefcase and, from several pages on a yellow pad, consulted a list of maybe eighty-seven or eighty-eight questions he wanted to pose about a potential partnership between his company and BSR."

Dunn remembers that after about twenty-five questions Warshaw seemed satisfied that BSR could be useful to him. They talked for more than two hours. At first Dunn felt that he was being interrogated. "Then Warshaw relaxed and I saw that he wanted to change the business and have it be run consistently with what he believed personally. We talked about the Lindner family. Warshaw hinted that if the public continued to perceive Carl Lindner as bad, then the company would be perceived that way too."

Beginning with the bewildering European trade system, the 1990s were a difficult decade for Chiquita. Perhaps this important meeting— and the year 1998 as a whole—was the moment in history when the wheels grinding slowly behind the scenes were first felt on the stage itself. The company crawled out from the strike in Panama and the storm wreckage in Honduras only to find the antidollar-banana forces in Europe in full cry behind the EU trade regime. The bitter aftertaste of the *Enquirer* episode did not evaporate quickly, either.

Despite this onslaught of troubles, Chiquita's managers believed they could ride out the storms. In 1998 the company still shipped millions of boxes of bananas from its other Latin American farms and continued its certification program with the Rainforest Alliance, at the same time pressing its case at the World Trade Organization against the European banana regime.

An *"Internetworked"* World

To understand the story of Chiquita over the past decade and a half is to understand many of the forces changing the way companies do business. The Chiquita case goes to the heart of key issues that arise as multinationals try to redefine themselves as financially successful and socially responsible businesses.

Global companies today are dealing with an "Internetworked" world, in which a movement of increasingly savvy consumers is merging with highly motivated environmental and social activists, and not just in Europe. Indeed, multinationals can no longer ignore the way the Internet has turned pockets of anti-American, anticorporate, antiestablishment rage around the world into a new force. From Seattle to Doha to Cancún, establishment institutions are getting a wake-up call. With their mission to level the global playing field for free trade, many fear that these institutions are running roughshod over workers and the environment, further deepening a global antipathy toward everything American—its power, its brands, and its culture.

With globalization, companies across all industries face a ratcheting up of public expectations and continuing mistrust of corporate motives. Activists are demanding mandatory codes of business conduct, and sophisticated campaigners are battling over environmental issues, workers' and women's rights, and corporate responsibility.

Building Trust

The list of charges raised by the *Enquirer* sounds very much like the chants you can hear on the street whenever rich institutions gather. Yet even the harshest critics are slowly admitting that Chiquita Brands International is running ahead of the pack. Alone among its competitors, the company had spent more than $20 million in the eight years starting in 1992 to certify the environmental performance of its owned farms in Central America, Colombia, and Ecuador. The Rainforest Alliance's Better Banana Program (BBP) of certification was the first in the agriculture business, and virtually unheard of in any industry.[3]

In 1992, when Chiquita undertook the pilot certifications with Better

Banana, multinational companies generally thought long and hard before they contracted with an outside party to review performance of any kind. In-house lawyers, especially in the United States, weighed such reviews with extreme caution lest they turn up something that could be used against the company by aggressive crusaders. Environmental audits by outside consultants were becoming more common, but the big engineering or management-consulting firms most often performed them under highly guarded conditions. Chiquita and the other banana companies had had a history of secrecy and lawyer dependency since their earliest days of doing business.

The challenges to the company culture must have been intense when Chiquita managers in Costa Rica actually permitted Rainforest Alliance operatives—surely environmentalists—inside the farms for the first time to certify that actual practice conformed to rigorous new principles. This revolutionary step called for a measure of mutual trust between the two organizations that still perplexes most observers. Since neither the Alliance nor Chiquita had any experience in on-the-ground certification, the gamble initially took an unexpectedly long time (eighteen hard months) to come to fruition. The certification process is rigorous, involving major changes at every step of the production process, and it questions agricultural practices ingrained over decades. Gradual reduction in overall agrichemical use is basic to the concept, and agrichemicals banned in the United States and Europe are forbidden. The overall purpose of the process was to address land degradation and prevent the spread of deforestation resulting from old-fashioned agricultural practice. It was to be a path to sustainable development that included protection of workers and community.

Assuming Social Responsibility

Since Chiquita was the largest employer of agriculture workers in Latin America, Steven Warshaw's initiation of a comprehensive program of social responsibility shook the industry again. The plan ultimately involved a close, independent scrutiny of company-wide labor practices. Warshaw said that Better Banana gave the company confidence to embark on an equally far-reaching innovation. "Better Banana is the granddaddy of Chiquita's whole social thrust," he said.

Sustainable Agriculture Certification Principles

Ecosystem Conservation: Farmers should promote the conservation and recuperation of ecosystems on and near the farm.

Wildlife Conservation: Concrete and constant measures must be taken to protect biodiversity, especially threatened and endangered species and their habitats.

Fair Treatment and Good Conditions for Workers: Agriculture should improve the well being and standard of living for farmers, workers and their families.

Community Relations: Farms must be "good neighbors" to nearby communities and a part of the economic and social development.

Complete, Integrated Management of Wastes: Farmers must reduce, reuse, and recycle wherever possible, and properly manage all wastes.

Integrated Pest Management: Farmers must learn to use nature and diversity as allies in crop production, rely on agrichemicals only as a last resort, and strictly control the use of any agrichemicals to protect the health and safety of workers, communities, and the environment.

Conservation of Water Resources: All pollution and contamination must be controlled, and waterways must be protected with vegetative barriers.

Soil Conservation: Erosion must be controlled, and soil health and fertility should be maintained and enriched where possible.

Planning and Monitoring: Agricultural activities should be planned, monitored, and evaluated considering economic, social, and environmental aspects.

("Although specific farm evaluation criteria vary by crop and regional ecosystem, all certifications abide by [these] nine principles." Rainforest Alliance Annual Report, 2001, p. 15. For a detailed description of the requirements under each principle, see www.rainforest-alliance.org.)

Starting at corporate headquarters, the social program was to be rolled out to the divisions and farms worldwide. It would be capped ultimately by independent third-party audits against the international labor standard Social Accountability 8000. Initial pilot audits were conducted by SGS (Société Gènèrale de Surveillance). (Social Accountability International, or SAI, the nonprofit human rights group that pioneered its model of social certification, in turn accredited SGS.) The results of Chiquita's internal reviews of performance, attested to by SGS, were published in Chiquita's first public report based on data gathered in 2000.

The BBP and SA 8000 certification programs have continued in the face of a worldwide banana glut and the financially disastrous aftermath of the protracted banana trade dispute between the United States and the European Union. Though finally settled, the trade war, along with some controversial management decisions, helped drive the company to bankruptcy in 2001. With the stock price at an all-time low, a reorganization, and new top leadership, Chiquita nevertheless insisted it would not abandon its risky social and environmental initiatives.

Reforming El Pulpo

One reason for Chiquita's tenacity is that these reforms may actually be putting to rest a century-old reputation that has proved very hard to shake. Indeed, Chiquita is the embodiment, a few times removed, of the storied United Fruit Company (known in Central America as El Pulpo), and is still identified in many minds with its nefarious banana republic image. In the empire years, United Fruit dominated the banana export business and exerted its influence throughout Latin America. The company controlled land, railroads, a huge fleet of refrigerated vessels, workers, governments, and even, it was said, certain members of the U.S. Congress.

Matters turned sour in the fifties, when El Pulpo was publicly implicated in the overthrow of the Guatemalan government. The anticommunist crusade was under way, masterminded by the U.S. State Department and the Central Intelligence Agency (CIA) under the Dulles brothers, John Foster and Allen. At the time, both the government and Chiquita profited from a slick promotion campaign led by the publicist

Edward Bernays. Though it was almost fifty years ago, this high-handed political maneuvering earned Chiquita a colonialist reputation that suited the rhetoric of its enemies perfectly.[4] Ironically, it ultimately served to trigger major reform in the banana industry.

The Guatemalan episode notwithstanding, United Fruit enjoyed a period of relative calm in the post–World War II years. Then in 1969, with the intention of becoming a real player in the food business, the corporate raider Eli Black bought 733,000 shares of United Fruit in the third-largest single-day transaction in Wall Street history. Now in control, Black renamed the company United Brands, cut research and development to the bone, and sold off a number of companies. All this changed when in 1973 he heaved his briefcase through a window of his forty-fourth floor of the old Pan Am Building in New York and dived after it, leaving behind tremendous losses. Black, who came up through accounting and finance, never really understood the culture or the problems that were at the heart of the banana business: the farms, the workers, the managers, and the governments in Central America. Although he volunteered humanitarian aid to hurricane sufferers, under his leadership United Brands bribed Honduran chief of state General Oswaldo Lopez Arellano with $2.5 million to reduce recently imposed export taxes.[5]

The Lindner Era

Eli Black's suicide triggered a reshuffling of ownership of the company. For a few years the New York real estate developers the Milstein brothers were in control. In 1976, Carl Lindner, one of the biggest investors in the company and one of the richest men in America, became president. By 1982, he had significantly increased his ownership stake and his family's control of Chiquita. His vehicle was a holding company, the American Financial Group, Inc. Insurance, including auto specialty lines and general commercial and personal coverage, is the company's main business. Principal subsidiaries are the Great American Insurance Company and the American Annuity Group. In 2000, Lindner also bought controlling interest in the Cincinnati Reds baseball team.

One hundred years after the start of United Fruit, in April 1999, the two U.S. senators from Ohio paid tribute to Lindner on his eightieth birthday. His wife, Edyth, and their sons, Carl III, Craig, and Keith, gave him a party attended by some five hundred guests, among them George H. W. and Barbara Bush and Donald Trump. The event featured Peter Duchin's orchestra and entertainment by Jay Leno and Frank Sinatra Jr. It was his son Keith whom Lindner had tapped to oversee Chiquita's operations.

When Lindner took over as chairman of the company in 1985, in the spirit of the times he streamlined operations, selling off soft drink, animal feed, and communications subsidiaries and appointing a new board of directors. Within a year, he moved company headquarters from New York to Cincinnati, the base of his other financial operations. Between 1985 and 1988 Lindner doubled the firm's cash flow. He expanded the Chiquita brand to grapefruit and pineapple, at the same time briefly recapturing first place in the global banana trade from Dole. In 1990 he changed the name of the company to Chiquita Brands International, a move to benefit the company as a whole from its best-known brand.

With the next management generation under Lindner's control and with a new top group of executives, a revitalized Chiquita was set to exploit the tremendous European Common Market and new markets in Eastern Europe and the former Soviet Union. The company geared up with a large-scale expansion of banana farms in Central America and expensive new shipping and processing facilities. The company soon learned, however, that in Europe it had to cope with its colonial burden and banana republic heritage. A major campaign was required to bolster its premium brand image and regain market dominance. The company had been building its Chiquita brand since the late forties. For the decade of the nineties, Chiquita ranked in the top ten of three hundred U.S. brands, according to Total Research, Inc., of Princeton, New Jersey. Of the top one hundred U.S. food companies ranked by sales in 1998, Chiquita stood at number 33 to rival Dole's place at number 22 and Del Monte at 55.[6]

Chiquita discovered the hard way that a highly visible brand can be a double-edged sword: while commanding a premium in the market, it can also be a lightning rod for critics. International labor unions, trade activists, and worker protection campaigners, especially in Europe, made no bones about the motivation behind their purposeful drive to discredit the oval blue-and-yellow label. With its shabby history and market dominance, Chiquita was—more than anything else—a symbolic enemy. Alistair Smith, head of Banana Link, the small-farmer advocate and Fair Trade proponent in the United Kingdom, admitted in our interview just after the launch of the EUROBAN anti-Chiquita campaign, "Unfortunately for Chiquita, they're the model that a bunch of reasonably well-organized, well-informed, well-educated, and multilingual people have chosen to use as a test."

That these campaigners were centered in Europe was a disaster for the company. Before the trade imbroglio, the lion's share of Chiquita's banana profits came from a European market willing to pay a premium for the Chiquita brand. As Gallagher and McWhirter reported in their 1998 *Enquirer* articles, Belgium's largest supermarket chain sold Chiquita bananas for about $1.10 a pound. (The average American price was about 50 cents a pound.) Another brand sold for 20 francs less, or about 85 cents a pound. "Shoppers—from elderly pensioners to teenagers—overwhelmingly reach for Chiquita."

By the 1990s the company was struggling, in part because of the protectionist European banana trade blockade and possibly also because of errors in business judgment, as some critics claimed. The company lost more than $700 million. Its stock price spiked in 1992 at $38.63, then dropped and bounced around at $10 to $15, finishing in December 1999 at $5.13. In 2001, Chiquita announced that it could not cover the interest on its debt of $862 million incurred during the big expansion; it filed for bankruptcy protection. A year later the restructured company, under the leadership of Cyrus Freidheim Jr. and managed by its creditors, had a clean balance sheet and was continuing to market a range of products: fresh fruits, juices, vegetables, and processed food products, along with bananas. Today the company owns or controls divisions in Africa, Europe, Australia, and the Philippines in addition to Latin America, and markets its products all over the world. (In 2003 it sold off its vegetable canning subsidiary to Seneca Foods Corporation for $110 million in cash and 968,000 shares of Seneca preferred stock. Chiquita will use the proceeds primarily to reduce debt.)

The Best-Laid Plans

Chiquita, along with its rivals, Dole and Del Monte, began preparing for a major expansion of banana production in the mid-1980s to reach the seemingly vast pent-up market unfolding in Eastern Europe and the former Soviet Union. It is said that the first food item the East Europeans asked for were bananas: fresh fruit had been a luxury for years, and bananas were a year-round bargain.

Then in 1994, after the companies had made significant investments in new farms, the European Union changed the rules of the export game. It installed a quota and licensing system for banana imports to give preference to those produced in its former colonies and territories in Africa, the Caribbean, and the Pacific (the so-called ACP countries). Chiquita, with limited investments in these countries, took its case to Washington. Senator Robert Dole, a continuous recipient of Lindner's fabled political largesse, proclaimed that the European initiative went against free market rules, and he proposed retaliation. Lindner's long-standing relationships in Washington enabled him to get the attention of the office of the U.S. trade representative and put the matter on top of the World Trade Organization (WTO) agenda.

The market for bananas in Eastern Europe and Russia had not materialized either. Between the trade-rule bottleneck and political and economic troubles in former Iron Curtain countries, thousands upon thousands of boxes of bananas that might have been going to Europe started coming to the United States at bargain prices. Meanwhile, activists began attacking Chiquita on every side.

Enter Better Banana

When Lindner bought controlling interest in Chiquita, he undoubtedly had no idea that sixteen years later, in partial response to this pressure, his company would have spent millions of dollars on independent "green" certification, with the goal of certifying 100 percent of its farms by 2001. In 1986, just two years after Lindner gained control of Chiquita, Cincinnati-born Daniel Katz, a 24-year-old China specialist for a Wall Street law firm, attended a small rainforest workshop in New York City. Strongly moved by the problem of rainforest loss described at the workshop, Katz convinced a group of dedicated volunteers to find a new way to address the problem. One early partner was Ivan Ussach, a toxicologist. Karin Kreider, a young video producer, offered her services soon afterward.

From the beginning, Katz was convinced that education and innovation were crucial in tackling a problem that had proved daunting to more traditional conservation approaches. But he also knew that he needed financial backers and he needed unquestioned scientific credibility. With dedication his most potent sales technique, Katz attracted interest and help from Ghillian Prance, a well-known botanist at New York Botanical Garden (now at Kew Botanic Gardens, U.K.) and the eminent biologist Thomas Lovejoy Jr. The former president of the Brentano bookstore chain, Leonard Schwartz, agreed to chair the board of directors. Schwartz was a serious collector of rare orchids and had experienced firsthand the reality of species loss that accompanies tropical forest destruction.

Katz and Schwartz secured a $40,000 grant from Paul Hawken (of Smith and Hawken) to start a program to certify green management of rainforests. Smith and Hawken at the time was buying teak from Indonesia. By century's end, Schwartz had died and Katz was running a $4 million conservation organization whose banana certification program had won the 1995 Peter F. Drucker award for nonprofit innovation.

While the Rainforest Alliance was certifying timber companies in Costa Rica and elsewhere under its "SmartWood" program, Chiquita had begun to buy land in Central America to convert to banana monocultures (much was degraded beef pastureland, but considerable first- and second-growth forest was also converted). Parts of a report commissioned by the Costa Rican government and leaked to the press stated that 3,000 hectares of forest were lost to bananas between 1990 and 1992. When the successor government took over, the report was shelved and the International Union for Conservation of Nature and Natural Resources (IUCN), withdrew its name from the report. American activists, already up in arms about the McDonald's "hamburger connection," now had a new enemy: the banana companies, especially the big three. Local and European groups took up the cry.

In early 1992 the Second International Water Tribunal, held in Amsterdam, blamed the Standard Banana Company (marketing bananas under the Dole label) for polluting Costa Rican rivers. The condemnation, brought by local groups and the prolabor organization Foro Emaús, was based on a study by the Association for Defense of Hydrographic Basins that had sampled water running from the Standard banana plantations farms over a fourteen-month period. The local English-language

Rangers: The ECO-OK banana project needs your help! You can write to some of the big banana companies. Tell them you hope they make all their plantations ECO-OK. Write to:

ECO-O.K.
Rainforest Alliance
A P P R O V E D

- Carl Lindner, CEO; Chiquita Brands; 000 E. 0th St.; Cincinnati, OH 00000.

- Gary Bratcher, CEO; Del Monte Fresh Produce Company; 000 Douglas Entrance, North Tower; Coral Gables, FL 00000.

- David Murdock, CEO; Dole Food Company; P.O. Box 0000; Westlake Village, CA 00000.

You can also tell the manager of your local supermarket that you want the store to get ECO-OK bananas.

To find out more about the ECO-OK project, you can write to Rainforest Alliance, New York, NY 00000.

CEOs of the Big Three banana companies received hundreds of letters from children. ECO-OK was the label used at the time. SOURCE: Reprinted from February 1994 issue of Ranger Rick *magazine, with permission of the publisher. Copyright © 1994 by the National Wildlife Federation. Addresses have been deleted.*

paper, the *Tico Times,* in July reported that thousands of dead fish had been found in the Tortuguero canals. Rocio Lopez, a biologist who headed the Association for the Environmental Well-Being of Sarapiquí (ABAS), blamed the disaster on the banana companies "because there's no other agricultural activity in this zone associated with the death of fish. We just don't have the proof."[7]

At about the same time, lawyers for fifteen hundred Costa Rican banana workers had won the right to sue the banana company Castle & Cooke and a number of large U.S. chemical companies in a Texas court for causing sterilization of the workers from chemical poisoning. The fledgling Rainforest Alliance was getting calls from a concerned public calling for a boycott of the big banana companies. Instead, Katz and his colleagues decided to approach the banana companies with a certification seal of approval that was beginning to be successful with tropical timber management under the organization's SmartWood label.

The Alliance's point man in Costa Rica at this time was Chris Wille (rhymes with chili). He collaborated with the National Wildlife Federation to put a center spread in *Ranger Rick* (February 1994), the federation's high-circulation nature magazine for young people. The spread showed a huge, innocent sea turtle being slowly strangled by the blue plastic bags then impregnated with insecticide and used regularly by all banana companies to protect the fruit and to ward off destructive insects. Kids were urged to write to Carl Lindner and the other banana bosses begging them to stop killing the turtles. When the letters began to pour into headquarters, "that was really a wake-up call," recalled Chiquita manager Dave McLaughlin. "All the banana companies say that," Wille told us.

Once the Alliance had decided to pursue banana certification, Katz wrote letters to the CEOs of the three big banana companies, explaining to them that his organization was different and offering to work with, not against, them. By coincidence Katz had played tennis in Cincinnati with Keith Lindner, Carl's son, when the two boys were in high school. After Carl moved his company headquarters to Cincinnati, Keith became the day-to-day Lindner at Chiquita. Katz recalled in our interview, "I wrote a letter to Keith, saying I hope you remember when I messed around with your tennis game; now I want to get involved with your banana company." Whatever the young Lindner's reaction, Katz's Cincinnati roots probably helped the Alliance get in the door.

Soon after the letter, Katz, together with Elizabeth Skinner (an early employee of the Alliance), journeyed to Cincinnati for the crucial first meeting between the two organizations. Katz recalls that there were about ten Chiquita people in the room, including banana boss Robert Kistinger and Magnes Welsh, the early go-between for Chiquita. There may have been at least one Chiquita lawyer present. Katz recalls the meeting as ending "somewhere between hostile and neutral." "There was one guy there who was all over us saying, 'We didn't know what we were talking about. Chiquita was doing nothing wrong on the farms,' etc. etc.," Katz told us.

But the door was left open for further meetings. For one thing, as Kistinger later said, "We had a problem and we knew we couldn't just pull something off the shelf. We looked around and the Rainforest Alliance was the best choice to work with us."

As the cautious strategic moves were playing out between Cincinnati and New York, more dicey tactical maneuvering was under way in banana lands in Costa Rica. Chiquita's Latin American farm management people agreed to pilot the Alliance's untried green certification program on two of their farms. Most people give credit for the early period of this risky new partnership to Chris Wille and Dave McLaughlin.

And risky it was, because virtually no company in any industry had ever opened its operations to inspection by an admittedly aggressive conservation organization, with no guarantee as to the results. McLaughlin certainly understood that ultimately Alliance certification in the agriculture sector would involve, in a pattern of continuous improvement, measurement of performance against commonly accepted rigorous standards for pesticide reduction, ground and surface water monitoring, creation of wildlife corridors, and worker handling of chemicals. These performance standards frequently involve expensive retrofitting of packing stations and other infrastructure facilities.

Both Wille and McLaughlin report that once the workers and the farm managers (often initially resistant as well) "get with the program," they become enthusiastic proponents. Chiquita can point to increased productivity and some incremental savings in expensive agrichemicals over the years. Warshaw said, "We couldn't prove in advance that there would be a return on investment, but I'm absolutely convinced it makes us more profitable."

Although Chiquita is alone in having its farms certified by Better Banana, the other dollar banana companies—fierce competitors always—have undertaken less risky in-house versions. For example, Dole and Del Monte have taken steps to become certified to the international environmental management system standard (ISO 14001), painstakingly developed by industry and modeled on the better-known ISO quality management standard—both under the aegis of the International Organization

Left: *A typical banana farm scene in the early 1990s, with a mountain of plastic waste dumped alongside a road. Floods would often carry this waste into rivers and eventually out to sea.* Right: *A drainage canal with effective soil protection. Before BBP, the canals were routinely scraped to keep them clean of vegetation, greatly accelerating erosion and the need for more canal maintenance. Changing this practice took years of research and convincing and training.*

for Standardization in Geneva, Switzerland. Under this system, third party auditors certify management systems. Dole claims to be the first agricultural company in the world to have all of its operations so certified. (For more detail, see Chapter 10.) Although it has been generally accepted by industry, environmentalists frequently dismiss the standard as being merely a paperwork approach that does not audit practice.

When Chiquita and the Rainforest Alliance negotiated the certification arrangement, it was not necessarily as partners or allies, yet they found themselves facing the same set of critics and enemies. These include proworker alliances mostly based in Europe, with some linked to various European labeling interests. Other critics included purist conservation groups that would rather boycott than certify and had raised questions about the objectivity of the Better Banana process. A notable criticism came from workers' rights advocates who claimed that Better Banana certification, even with its spot checks and annual recertifications, could not possibly verify that Chiquita had a day-to-day spotless record among its thousands of banana workers.

Not Invented Here

Among the handful of truly well informed critics during the campaign of the 1990s, Alistair Smith and his Banana Link group in the United Kingdom were the most tenacious. But like protesters worldwide these days, not all critics have motives of unalloyed virtue. As we will see, some of the most vocal protesters, especially in northern Europe, get funding from their governments for projects in Latin America.[8] Others are simply business rivals in various garbs, who gain when Chiquita loses. The unavoidable contest between BBP and the European labeling schemes of Fair Trade and organic, and the U.K. Ethical Trading Initiative (ETI) was heating up. Oxfam and Christian Aid, among others, financed certain of those interests.

Even with the important role of the ICCR, the Sisters of Charity, and the workers' advocacy group U.S./LEAP, it is impossible to ignore the growing number of critics who simply do not like the United States of America. The not-so-faint whiff of "not invented here" is inescapable. Chiquita's critics had United Fruit's long unsavory history in Central America to pummel. The Stars and Stripes flying over vast banana plantations was just too provocative an image to pass up.

3

Times Change

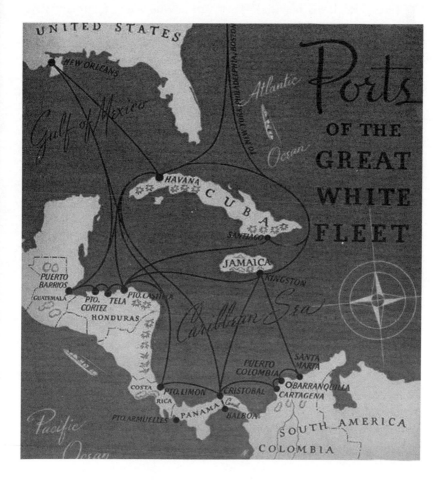

The world loves bananas. They are a vital part of human diets throughout the tropics. Farmers can grow 44,000 pounds of bananas on a piece of land that can produce only 33 pounds of grain or 98 pounds of white potatoes. Almost 70 percent of American households eat bananas at least once a week. (Less than half that number eat apples with the same frequency.) The average U.S. shopper buys 28 pounds every year—more than any other fresh fruit. Some estimates maintain that German and Swedish families consume up to 40 pounds of bananas every year. Unlike other food commodities, bananas are in season every season. In the developing world, the banana and its greener cousin the plantain provide more than one quarter of all food calories to people in many parts of Africa.

And bananas seem to love the world: they almost overwhelm it with new progeny every day. Bananas are grown in all tropical regions and are the world's fourth most important food crop after rice, wheat, and maize in terms of gross value of production.

Bananas are both a staple food and an export commodity that represents a significant source of income and employment for several developing countries. The world's most exported fruit in terms of volume, they rank second after citrus in terms of total value, and first in terms of export revenues to developing countries. International trade in bananas is valued at close to U.S. $5 billion per year. Preliminary results from a model developed for the Food and Agriculture Organization (FAO) indicate that the world exports of bananas will reach some 13.9 million tons in 2010. Most of the growth predicted by the model will be in Latin American countries, specifically in Ecuador and to a lesser extent in Costa Rica and Colombia.[1]

As Woody Allen can testify, bananas are also just plain humorous. For some reason, a person (especially a solemn person) slipping on a banana peel is much funnier than a pratfall from slipping on anything else. And, let's face it, the size and shape of the banana make it downright salacious. The famous etiquette columnist Amy Vanderbilt once recommended to ladies that, except at picnics, bananas should be peeled and then broken up "as needed into small pieces" and "conveyed to the mouth with the fingers."[2] To many with long memories, the banana is associated with peril. Exotic fauna often accompanied the bananas when they were still shipped on the stalk. A retired employee of one of the companies told us that in the old days his job was to stay in the hold of the ship for the entire voyage and shoot tarantulas as they scuttled among the bunches.

Not only are bananas prolific, funny, sexy, and scary, they are also cheap and healthy. The fruit is rich in carbohydrates, vitamin C (also A and some B), and important minerals including potassium, copper, magnesium, calcium, and iron. In Uganda working males consume five pounds of banana pulp per person per day, equal to about 3,600 calories. Bananas have the same caloric value as white potatoes, with no cholesterol. Since they also offer a low level of sodium, they are appropriate for salt-free diets. Contrary to popular belief, bananas are low in fat. They have become a staple in newspaper health columns as a cheap and natural cure for everyday ailments. The zip-lock skin protects the fruit and makes it easy and safe to eat.

Ups and Downs

For the bold player, in the right place at the right time with deep enough pockets, such a treasured crop would seem to promise a bonanza. But the banana game—especially the way it is played by the major multinationals—is a complicated and exceptionally unforgiving business. As

What does a banana have to be to be a Chiquita?

It's sort of like passing the physical to become a Marine.

The banana's got to be the right height. The right weight. The right everything.

Right off, it has to be a good eight inches along the outer curve. And at least one and a quarter inches across the middle. It has to be plump. The peel has to fit tightly. The banana has to be sleek and firm.

It has to be good enough to get through a 15-point inspection. Not once, but three separate times.

Occasionally, though, a banana comes along that's got everything going for it—except maybe it's a smidgen under minimum length.

We should pull it, we know. But our inspectors have hearts, too.

Which is why you may sometimes find a Chiquita Brand Banana that isn't quite eight inches long.

Come to think of it, though, you sometimes find Marines named "Shorty," too.

Chiquita Brand Bananas.

A Chiquita advertisement in the late fifties. After reading the text, Life *magazine rejected the ad, although several women's magazines allowed its publication, according to McCann,* An American Company, *p. 87.*

one senior executive says, "It's like golf: the drive is the show but what counts is the putt." The most dangerous part of the trip from plantation to produce section for this fragile fruit is the last 30 feet at the supermarket, where an errant cart maneuvered by a careless stock clerk can undo thousands of tiny careful protection measures and spoil the single most profitable item in the produce section. In the jargon of the trade, the bananas could lose "one whole quality point."

The coddling to ensure that each bunch reaches the produce section at the highest possible peak of perfection is an expensive step-by-step process that the industry has inflicted on itself over the years. The stopwatch starts when the banana is harvested, and it ticks inexorably over road, ocean, rail, and ripener. No more than twenty-five days can elapse from farm to market. The pampering that takes place—from the pads inserted between the individual "fingers" at harvest to the delicate handling of each bunch in the packing stations on the farms, to the smoothing of farm roads, to the stacking on the docks, to the rapid loading and unloading of the refrigerator vessels, to the fine-tuning of color at the ripening centers, and so on right to the supermarket—is a process shoppers take for granted, but they are probably the only people in the entire banana chain who do.

With so much fruit and such fluctuating profit margins, the task of matching short-term variations in demand to so many different geographic markets with an expensive stream of supply (totally dependent on the vagaries of weather and labor) is a daunting exercise in logistics. One veteran of the business likened it to "chess on the high seas."

Very few products—food or otherwise—have such extreme price volatility at retail. If you track the price of a pound of bananas every day over a year, the swings can be as high as 300 percent. It is perfectly possible that at a given moment in time it may cost more to produce a box of bananas in Central America than it can be sold for in Germany a week or more later. Because of this volatility, bananas are an investment for shareholders with nerve and staying power. But because bananas are a very popular year-round tropical fruit, they are the great lead products in supermarkets, accounting for about a third of all produce sales. If an entrepreneur can stay in control and has sufficiently deep pockets, millions of dollars can be made in the banana export business.

Technical glitches can be crucial. In one case reported in the *Wall*

Street Journal, a Del Monte refrigerator ship lost power and a $1.4 million boatload of the fruit filled the decks with a mushy fragrant sludge. The experience would have been eerily familiar to Captain Lorenzo Dow Baker at the end of the nineteenth century, when the seagoing entrepreneur brought bananas to the States from Jamaica by sailing vessel. Becalmed in the tropics, he was forced more than once to dump his precious cargo overboard.

The banana business on the scale practiced now by only a handful of companies, all with decades of experience, carries with it an almost prohibitively high cost of entry. Along with the "dollar companies" (Chiquita, Dole, and Del Monte), Noboa (a huge family enterprise that ships its fruit out of Ecuador) and the Irish firm Fyffes (largely dependent on bananas from the Caribbean) dominate. Although there are many small and medium-sized independent producers, and while Australia, Africa, and the Philippines are producing an increasing number of bananas, most of the farms either are owned by the big companies or supply the fruit to them. There are literally no new players.

The challenges of production are daunting. To make money from the world's huge banana appetite, a successful company must not only cope with hurricanes, drought, flooding, the leaf fungus disease Black Sigatoka, and Panama disease (a vascular wilt disease involving the fungus *Fusarium oxysporum*), but also man-made hurdles like host governments with unpredictable political agendas; rigorous self-imposed quality constraints focused on size, color, weight, and curvature; rigid ripening schedules; arcane trade restrictions; labor disputes; price-cutting competition that verges on dumping; and currency fluctuations.

Moreover, any company operating a global business over decades, especially where the sources of supply are in developing countries, must more than ever be able to cope with huge culture gaps, and increasingly with social values and expectations. These expectations for environmental and human rights performance, most prominent in Europe and the United States, have put companies on the wrong websites in minutes. To McDonald's and Monsanto, and a mounting list of other multinationals, this possibility has become much more than a trivial public relations problem.

Banana Definitions

Finger - banana

Hand - group of bananas on a substem connected to the main stem. Bananas in markets are usually displayed in partial hands. Most hands are cut in the packing station because the ones in the middle are larger than those on the ends, and bananas are packed according to size.

Bunch - the entire group of fruit on a plant, all grown from a single, central stem.

Stalk - alternatively, stem. Traditionally, the stalks were dumped in piles to rot—often along streams, which they polluted. Under BBP standards, the stalks must be either returned to the field as natural fertilizers (not always possible because they can aggravate some diseases) or sent to managed, on-farm landfills.

Stem - most often the main stem of the whole plant, the "trunk." Bananas are a forb (a broad-leaved herb), not a tree, so "stem" is the right word. "Stem" and "stalk" are often interchanged. Sometimes "bunch" and "stalk" are used interchangeably. SOURCE: Chris Wille, Rainforest Alliance.

The Golden Years

Like the Dutch East India Company before it, United Fruit, in its earliest adventuring in Central America, imported most of its basic values from home.[3] In the entire period from the turn of the century literally into the 1950s, the company maintained an "enclave" culture, much of it carried over from the southern slave society. It was simply transferred farther south into the isthmus as slavery was abolished in the United States. The agricultural managers were mostly Americans from southern towns and farms who brought their old ways with them. They treated the dark-skinned workers with a peculiarly southern mixture of patrician civility and arrogant condescension. Until the 1930s most banana workers were imported into Central America from the West Indies. Overseers from Louisiana and elsewhere in the Deep South who ran the tropical farms believed these workers to be more tractable than others, and being native to former British colonies, they spoke English. Further, the workers, isolated in enclaves, were more manageable than the local subsistence farmers, who strongly resisted giving up their small plots to labor for the Yankee dollar in coastal company towns.

Banana production was jarring physical work. At harvest time, sheer muscle was called for; one stem—the "trunk" holding the entire group—of the bigger bananas marketed to the United States could weigh nearly 90 pounds. A worker would climb up to cut the banana bunch. It would fall a few feet onto the bent back of a worker waiting below, who would then jog the heavy burden to a nearby railroad car or be relieved by a second runner, who would take it to a distant mule cart.[4] Twelve hours a day of this physically demanding work made back injuries inevitable. Workers also suffered wounds from the machetes used to hack through the jungle, mucky and slippery from the tropical rains. The sharp knives needed to separate the bunches from the tree were another hazard. Snakebites were common, and so were spider bites. Little thought was given to the viability of the ecosystems or to the impact on worker health of the palette of chemicals used to ward off fungi and pests. Nor did the pampering that every banana enjoys today exist at that time.

Jungle Fiefdoms

As we enter the current century, the harmful impacts of banana operations on workers and the environment are the lingua franca of scores of nongovernmental organizations around the world. However, the image of a banana farm may be somewhat dated. In the mid-1950s, Thomas McCann, a former executive of United Fruit, could paint a graphic picture from his earliest days as a public relations employee:

> There was something at once very exciting, very masculine and very romantic about the company . . . It was a mixture of John Wayne movie clichés and the legacy of an incredible period in history; gin and tonics and Dewar's White Label scotch on tropical verandas; endless miles of private jungle fiefdoms; natives who were variously brooding, surly or submissive; boots, khaki uniforms, horses and

pistols; the Great White Fleet that was really the largest private navy in the world (and operated on the open-secret motto "Every banana a guest, every passenger a pest"); the early morning produce markets and the colorful, crude men who ran them; longshoremen, traders, plantation managers, ambitious men, hard men, lazy men, rich men; and behind it all a tradition of enormous wealth and power and privilege that was already starting to decay.[5]

During this era the hierarchical arrangement of a United Fruit plantation was symmetrical: the best address was the house on the hill (sometimes artificially built up) of the division manager. Subordinates' houses were just a little lower. The divisions were set up like army camps. Along with the company stores, most divisions had three classes of clubs: the top club was the same as an army officer's club; next in the pecking order was the second-class club for the Latin clerical and supervisory personnel (in effect, the noncommissioned officers); at the bottom were the clubs for the workers—the privates and corporals.

The so-called American Club was almost always on a golf course, world class and manicured by cheap local labor—just for the United Fruit executives. Famous golf pros were frequent visitors. Multilane bowling was the preferred activity in the second-class clubs. The third-class clubs were the sites of all the real action. Unless one felt truly adventurous and crossed over into the "civilian quarters," with poverty everywhere, cantinas, whorehouses, and few friendly faces.

In 1952, the time McCann was writing about, United Fruit had fully a third of the world's commerce in bananas. More than half of that trade was in the United States and Canada, where the per capita consumption was 17 pounds a year. The company had liquid assets of more than $50 million and sixty-six thousand shareholders. Its stock was selling for $65 a share. It imported 30 million stems of bananas per year into the United States and Canada and some 5 million into Europe.[6]

For division managers and their wives, life in the tropics was so extraordinary that a transfer stateside was contemplated with fear and trembling. Where in the United States could a division manager's wife enjoy a status equal to that of the wife of the president of the host country? These families lived rent free in the best house in town. They had access to semiprivate medical services. Their entertainment expenses were unlimited. Of course there were servants. Transportation to anywhere in the world was courtesy of either the Great White Fleet or a twin-engine, radar-equipped airplane. Beach houses were not uncommon.

Much in United Fruit's way of doing business was the dark underside of colonialism. Aspects of the company's paternalism were obviously driven by the need to pacify the workers. In Guatemala, for example, the company introduced welfare policies for its workers that previously were unknown in the country. These included housing for the plantation workers, health care, hospitals and clinics, and schools for the worker's children—all at company expense.[7] United Fruit also provided money and effort to protect the Mayan anthropological treasures in the country, measures that not even the Guatemalan government had thought to undertake.

In Honduras, the company founded the Escuela Agricola Panamericana, an independent school of tropical agriculture, and created the Lancetilla Botanical Gardens, which still holds a major collection of tropical plants. The company built and staffed research stations to seek remedies for tropical diseases.

United Fruit was a very big landowner in Latin America in the 1950s. A scholarly work by Stacy May and Galo Plaza provides valuable perspective. Using figures supplied in 1955 by the company, the authors note that United Fruit that year owned 1,726,000 acres of land (2,700 square miles) in Guatemala, Honduras, Costa Rica, Panama, Colombia, and Ecuador, with the largest holding in Costa Rica at 35,000 acres. Somewhat less than a third of the land owned in the six countries was actually planted in bananas. That year the company's 388,000 acres of land planted in bananas and other crops was a little over 2 percent of all reported croplands in the six nations combined.[8]

McCann and a host of other critics have claimed that the company held so much excess land to prevent competitors from moving in. No doubt there is some truth to this statement, and the charge was raised in the 1955 antitrust suit brought by the U.S. Department of Justice. May and Plaza offer additional reasons for this vast acreage. First, the substantial infrastructure of houses, roads, hospitals, churches, schools, railroads, and the like requires more land than the area where the bananas actually are grown. Second, low-lying cropland requires a sizable network of drainage canals. Third, the fight against Panama disease involved a process called flood fallowing, in which massive dikes were built around each farm in an infected area; then water was released into the enclosed areas for periods of up to several months. Even this expensive control measure failed quite often, in which case farms were simply abandoned. May and Plaza cite the company's own records, which specify that over the fifty years prior to 1954 cultivation was discontinued on some 900,000 acres of land, mostly in retreat from Panama disease. Finally, in acquiring land, the company frequently had to deal with landowners unwilling to sell only a portion of their properties.

Still, in the 1950s and even later, the tropical forest was dismissed as a mere jungle: modern conservationists and small-farmer advocates would be struck by May and Plaza's perspective that there was little demand for this "nonbanana land" because it was generally surrounded by "countless acres" of entirely comparable tropical forests being put to no productive use. Perhaps the banana overlord's ears continued to ring with God's biblical injunction to Adam and Eve: "Be fruitful, and multiply, and replenish the Earth and subdue it."

Long-Distance Management

From United Fruit's earliest days, first in Boston, then in New York, and now in Cincinnati, the common wisdom has been that the top brass count and allocate the money at a safe distance from the scorching, muddy reality of the banana business on the ground. Yet strategic decisions made in the corporate suites often had dramatic effects on the farms. It was at headquarters that the fundamental decision was made in the 1960s to replant all the farms with the new, less disease susceptible banana variety Valery, a clone of the Cavendish type.[9] Another decisive moment came when headquarters exploited mass marketing techniques to transform what is, after all, an agricultural commodity into a branding strategy, and to bolster that novel branding approach with a rigorous set of standards for quality.

These innovations made up for lost time and set new standards for the industry. The idea of branding bananas was not new in the midsixties, when Chiquita CEO Tom Sunderland hired John M. Fox from Minute Maid Orange Juice, following his brief stint at Coca-Cola when Coke bought Minute Maid. It was Fox's idea, in the face of scorn from managers in the tropics, to exploit the popular Chiquita image (at the time the property of United Fruit, but used generically by all companies) by putting the familiar cha-cha Latina figure directly on United Fruit bananas. After much experimenting, a young Honduran laborer rigged a device that power-fed labels onto bananas with just an operator's thumb and forefinger. With that, the company became the largest single user of pressure-sensitive labels in the world—and the banana industry took one more step in the uphill struggle to convince consumers that one banana really is different from another banana.

Having taken this bold step, Fox as chairman knew he had to back up the brand with a painstaking scheme for quality control. This crucial decision marked the earliest move in the industry to change what was essentially a commodity to a popular tropical fruit whose appearance, at least, could be fine-tuned to customer preference and for which a premium price could be charged.

The European Campaign

In 1967 the company launched a full-scale marketing assault on the European market to support the new Chiquita label that it was now putting on the Cavendish/Valery variety. The Gros Michel variety had been the industry choice until its susceptibility to disease made it untenable for large-scale production. It had been marketed in Europe by United Fruit under the Fyffes label. The year 1967 was also when Del Monte became a player in the international market with its acquisition of the West Indies Fruit Company.[10] That same year Castle & Cooke bought the Standard Fruit and Steamship Company. United Fruit's principal dollar competitors were on the march.

Chairman Fox's aggressive new campaign was later to test the authority of the European Economic Community (EEC) to challenge foreign firms if their actions threatened to overwhelm companies in member countries. It would also test the mettle of its mortal rival, Standard Fruit, which was selling in Europe under the Dole label. Ultimately, the campaign and the ensuing reaction against United Fruit were a foretaste of the banana battles of the 1990s in which the Rainforest Alliance found itself embroiled.

At the launch of its marketing push in 1967, United Fruit invited a number of companies throughout the EEC to distribute and ripen Chiquita bananas. Determined to back the Chiquita brand with uniform high quality that would enable the company to charge a premium in stores, United Fruit took extraordinary precautions to prevent its green bananas from losing their quality edge even after the official transfer of ownership had taken place. For example, in order to do business with the company, the new European distributors had to have ripening facilities of a very high technical standard. It was not unusual for the company to loan money to these firms, enabling them to build or modernize their ripening facilities. Quality control included the generation of an advanced technical assistance and control department. Because of this capability the company was able to advise ripeners, coordinate logistics, set ripening standards, train personnel, and follow up with spot checks.

In the late sixties the marketing campaign in Europe was in full swing. It was in 1970 that Eli Black took over United Fruit, renaming it United Brands and merging it with AMK Corporation, a major meat producer. During that period, United Brands had interests in flowers, palm oil, soybeans, rice, peanuts, and vegetables of various kinds, and in a wide range of preserved foods using these products. Following the restrictions imposed on United Brands by the American antitrust authorities, production, transport, distribution, and marketing of bananas worldwide in 1973 represented only 75 percent of the company's gross profits of $2 billion.

By 1974 United Brands had more than 60,000 employees, owned more than 30,000 hectares of banana plantations, and sold more than 100 million boxes of bananas, representing some 2 million metric tons, or about 35 percent of world exports. It was the largest banana company in the world. It also owned one of the world's largest banana-boat fleets and in addition used one of the most powerful refrigerator shipping operators, the Swedish firm Salén Shipping Company.

Salén was a large shareholder of United Brands. A subsidiary, United Brands Continental BV (Rotterdam), was responsible for coordinating the company's banana operations in all member states of the European Economic Community except Italy and the United Kingdom. In addition to its Latin American holdings, United Brands bought fruit from independent producers, including virtually the entire production of Surinam, Cameroon, and Guyana and a large proportion of the bananas

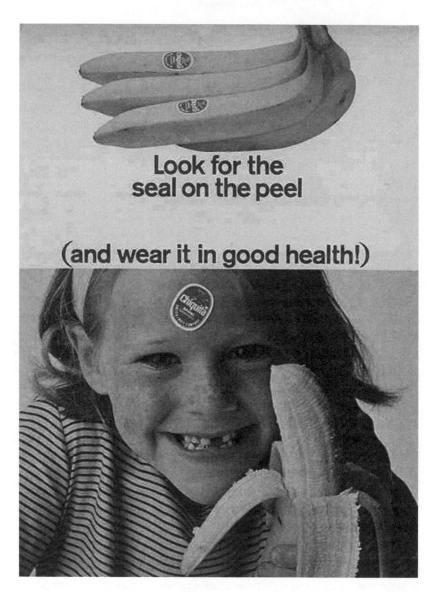

A Chiquita ad circa 1970

grown in Jamaica, Guadeloupe, the Philippines, Ecuador, and the Dominican Republic. The company's main port of delivery in the EEC was Bremerhaven. Most of the bananas for sale in Germany, Denmark, Switzerland, and Austria passed through this port. Its second main port was Rotterdam, from which bananas headed for the Netherlands, Belgium, Luxembourg, Germany, Ireland, and sometimes Denmark.

The company's marketing policy in Europe insisted that all fruit be displayed with a Chiquita label, whether sold by its own subsidiaries or by other distributors/ripeners. Consumer attachment to this brand was sounded out and, if possible, strengthened by in-store promotion and advertising. Retailers and supermarkets apparently attached great importance to these efforts. United Brands coordinated everything with distributors and ripeners. Sometimes company representatives backed the promotions with cash payments to the store during the campaigns. For a time, United Brands was the only company involving the retailer so fully in stressing the general superiority of a banana with a Chiquita label affixed. In Chiquita's advertising, the company emphasized the value of a finely graded, high-quality fruit branded in the producer country. It assured retailers of steady supplies of bananas of uniform quality and outstanding appearance, with delivery timed to ensure the longest possible shelf life for such a perishable fruit. In essence, it provided a "zero defects" guarantee. Any bananas not satisfying these strict criteria could not carry the Chiquita label and were sold without a brand name. United Brands was able to do this because it could supply very large quantities of bananas of uniform quality. According to the information supplied by United Brands to the EEC, this marketing policy enabled it to sell Chiquita bananas at a price, on average, 30 to 40 percent higher than its unbranded fruit.[11]

The Banana Game

Branding is the name of the game, and it continues to be central to Chiquita's marketing strategy today. Sherrie L. Terry, Chiquita's vice president of marketing for Chiquita Fresh North America, told us, "Our mission is to create a relationship that promotes eating right with a brand that is fun and exciting in ways that only Chiquita can offer." Terry and Robert F. Kistinger, president and chief operating officer of the Chiquita Fresh Group, agree that the Chiquita brand is the single most valuable asset the company owns—"more valuable than our farms and more valuable than our ships," Kistinger says.

This deep belief in the strategic role of branding goes some way toward explaining senior management's decision to pursue environmental certification. In his office in Cincinnati, Kistinger did not hesitate to talk about the strategic purpose of BBP. A lean and intense figure, he thumped his hand on the table in his office and spoke rapidly: "We believed that it was critical for Chiquita to take a leadership position on environmental matters because Chiquita has always been a leader, and we needed to be a leader in this. Our whole certification decision was totally customer and consumer driven. That's the bottom line. The brand is our most critical asset. We don't want to do anything that will jeopardize this brand." Picking up a letter that had been on his desk from the Association of National Advertisers in New York, he pointed to a phrase he had underlined: " 'On average the brand accounts for 40 percent of corporate value.' Certification is not something we were forced to do. We have known for a long time that food safety is critical in the minds of the customer." (It is industry practice to draw a distinction between the customer, that is, the distributor/ripener or supermarket, and the ultimate consumer.)

Still, from the farm to the brand, and from United Fruit to Chiquita Brands International, the road has been long. To understand the scope and complexity of the banana industry, think of the board game Monopoly. You learn one lesson early: controlling property is great, but you can get zinged by the costs of that ownership.

The game is not a bad way to grasp the crucial differences between Chiquita Brands and its rivals: all those houses, all those hotels, and even when you draw an occasional "Get out of jail free" card, there are all those surprises from Chance and Community Chest. Even with the antitrust divestitures of the fifties and sixties, and with the selling of much of its land during tough times, Chiquita owns far more land in Latin America than its rivals, Dole and Del Monte. Owning your source of supply can be a huge advantage, eliminating many of the vagaries that ensue when another owner is involved. When you have pioneered an organizational structure in which you control not only the means of production, but almost the entire chain of custody, your options multiply enormously. When times are favorable, this ownership arrangement can help you hold a dominant place among your rivals, as was the case for decades with United Fruit. Unlike the Monopoly game, owning farms in several countries can spread the risk from unpredictable governments and predictably ruthless tropical weather. When times are not so favorable, ownership of land, workers' houses, and clinics, although these are still assets on the books, carries some strong obligations. It incurs not only financial expense, but also responsibility that goes beyond money. All that land can mean red ink when an unexpected shortfall in demand enters the picture—as it did when the European Union lowered the boom with its protectionist trade regime just when Chiquita had bankrolled a vast expansion in farm acreage.

Owned farms also mean a huge payroll to meet and union demands to negotiate. "We are the largest employer of unionized workers in the agricultural sector in Central America," said Kistinger. One can begin to see how soaring international expectations for corporate social responsibility began to loom larger and larger in the strategic thinking of the company. United Fruit's long paternalistic past forced a burden on the modern Chiquita, with immense political ramifications. "Nobody appreciates what a juggling act the banana business is," he added. "Think of one giant open-air factory." In addition, the company has had to deal with its hundred-year history. "The world is not the same as it was a hundred years ago. When we were first doing business as United Fruit, we were major contributors to the social and economic development of a number of Latin American countries."

Asked to explain the logic of putting management personnel at risk in the war zones of Colombia, especially in the face of an acknowledged global glut of the fruit, Kistinger replied that bananas are still a social business. "In a place like Colombia, God help them if we went out of business." It would be a disaster. Hundreds of small operators sell to Chiquita, most with soft loans from the banks. The banks, and everyone, would be in danger of going under. "We are responsible for social stability in a very large region," Kistinger remarked, the set expression on his face revealing his sharp sense of commitment and responsibility. Certainly the benefit of social stability has not been lost on the governments in any of the exporting countries, but it is this very dependency that is at the heart of the anticorporate position.

In a nutshell, farm ownership almost always means employees, and after decades of resisting, Chiquita has accepted that employees mean unions, and that these become a major competitive factor. As the unions successfully bargain for higher wages, productivity emerges as a central issue. A senior Chiquita executive tells the following story: He was interested in beefing up productivity in Panama, where dealings with banana unions have historically been difficult for the company. He sent down a skilled worker from Costa Rica to show how bananas could be harvested more efficiently. "They took one look at this guy and they told me I had sent down the Michael Jordan of banana cutters." This anecdote speaks volumes about the essence of labor/management strife.

Labor is typically 35 percent of the cost of a box of bananas. The competitive advantage is enormous when Noboa and Dole buy most of their fruit from independent farmers in Ecuador who are paid about two dollars a day, and Chiquita's independent supplier in Ecuador, Reybancorp, pays its workers at least the legal minimum wage guaranteed through BBP certification. Add to that the fact that most environmental measures are ignored on the small, usually cash-poor Ecuadorian farms. It all exacts a hefty short-term penalty on Chiquita for its decision to be environmentally and socially responsible.[12] Only recently have workers' rights activists broken their long silence about the labor practices and environmental price paid as a consequence of cheap Ecuadorian bananas. U.S./LEAP, the Chicago-based labor advocate that had been one of the few U.S. critics focused on abuses on Latin American banana farms, was the first to swing its beam onto the issues in Ecuador. Human Rights Watch and Banana Link (U.K.) joined in to turn up the heat, especially against Noboa.

For any company to survive more than a hundred years is something like a miracle. A close inspection of the Fortune 500 list over time tells the story. The survivors have learned how to adapt to seismic cultural changes in technology and markets. But as Philip Morris has learned, times change. In reflecting on these changes, Rivaldo Oyuela Martinez, a long-time quality control engineer for Chiquita in Honduras, said recently that he remembers the neighborhood parties and the Chiquita executives playing golf on their private course. Now most members of the golf clubs are *maquila* executives from Japan and the Philippines.[13]

The impressive changes Chiquita has made to adapt its brand to a market consisting of increasingly political consumers, especially in Europe, have helped put it ahead of the curve. Staying there will depend on much more than certification alone.

4

Why Bananas and Why Chiquita?

It is difficult today to appreciate the virulent opposition confronting Chiquita in the 1990s, even though the larger conflict between advocates of small-scale agriculture and single-crop industrial agriculture will be with us for a long time. During the nineties social activists in Europe singled out the big multinational banana companies as the symbol of all that was wrong with the big multinational agricultural companies. Their number one target was Chiquita, and the leading campaign organization was Banana Link, which has its headquarters in Norwich, England.

Although itself quite small, Banana Link has cast a long shadow in the banana industry. In a few short years it succeeded in bringing together a diverse network of European NGOs and trade unions to form the European Banana Action Network or EUROBAN. In addition to COLSIBA (forty banana workers' unions in eight countries of Latin America), EU-ROBAN's main partners across the Atlantic are the Windward Islands Farmers' Association, Foro Emaús (a Catholic militant workers' rights group in Costa Rica), and the U.S. Labor Education in the Americas Project (U.S./LEAP). This effort was linked from the outset with Fair Trade, the commercial labeling venture tied to retail shops in the U.K. and elsewhere in Europe that promotes equitable treatment of small farmers and entrepreneurs. The idea was to create a network competitive with the big banana exporters, which promised to reward growers in the Third World with a greater share of export revenues in contrast with the free-trade, multinational company model preferred by rich-country governments.

In 1998 EUROBAN, in concert with Gilberth Bermudez Umaña, the secretary general of the Costa Rican trade union SITRAP and deputy co-ordinator of COLSIBA (the Coordinating Latin American Banana Workers' Unions), convened a large conference of the warring parties in Brussels. The rules of engagement for the foreseeable future were defined, constituting one of the blows Chiquita faced that year. The conference was a bold effort to address what many had come to believe was a crisis in worldwide banana agriculture, with so few multinationals controlling such a huge share of the banana export business, devastating workers and the environment in the process. In addition to the NGOs and a number of academics specializing in agriculture, participants included government and company representatives. White papers and denouncements were circulated; the dollar banana companies came in for a drubbing, chief among them Chiquita. Why bananas and why Chiquita?

"We're not going to go away," Alistair Smith, Banana Link's executive director, told us in a clipped British accent one afternoon in Paris. "And neither are the big banana companies. Ultimately, we both want a quiet life, but our mission is to try and change things. We don't have a problem with multinational companies per se, but we do have a problem with companies that abuse their power toward people who work for them, and the environment." He went on to say: "Unfortunately for Chiquita and the others, they're the model sector. Just because they've started to change we're not going to roll over and say 'yeah, you're lovely, do whatever you want to do now.'"

Smith was in Paris lobbying trade and foreign affairs bureaucrats to see things Banana Link's way. The World Trade Organization meeting in Seattle was only a couple of months away. We sat down with Smith in the dark basement lounge of a small Left Bank hotel to listen to his story. Skeptical and cautious at first, he relaxed a little as the afternoon wore on.

Dragging deeply on his cigarette, the lanky former linguist, now small-farmer advocate, was a very plausible modern activist. Speaking with modest pride of having contributed to the recent changes in Chiquita's behavior, he said, "We didn't select Chiquita or the banana sector innocently or by chance. It was a strategic choice. The banana is a symbolic commodity at all sorts of levels. Soy beans just aren't as sexy, even though the basic story is the same—vast unsustainable monocultures controlled by a handful of multinational companies."

In part, Smith was reflecting a European view about a crisis in the relation of farming to the overall culture. Many Europeans strongly prefer the intricate patchwork of field, forest, and town to the extensive single-crop monocultures that prevail in much of the North American Midwest and in parts of Latin America and Asia, and are now threatening to change the European landscape. They see banana farming, as practiced by the multinationals in Latin America, as not only ecologically destructive but also culturally barren and demeaning to small farmers. Europeans, especially the French, understandably value their famous landscape, and the farm subsidies that help support the small-farm culture are increasingly being threatened by countries wishing to export their produce to Europe.[1]

EUROBAN got the attention of the big companies and banana wholesalers in Europe. At this point Smith and a dedicated group of crusaders had been campaigning in defense of agricultural workers for six years. Smith is clearly proud of his accomplishment. Describing the secret of his organization's success, he said that groups such as his can effect important changes in the way companies operate only if they have a deep understanding of how power works and a vision of what the future could look like. He believes there is an urgent obligation to fight apathy everywhere. In the United States, and even in his own country, the sense of disempowerment is overwhelming. "But that won't wash. We live in apocalyptic times. Bananas are a symbol of the madness and the hope. If you knew the kind of violence, the threats of violence—yes, even death threats . . ." His voice dropped off, "All of us have come close to the edge at various places and times in this process."

Banana Colonizers

All three of the big dollar-banana companies have been exporting from Latin America for decades. The long and checkered history of Chiquita, with its origins as United Fruit, made it the ripest target for EUROBAN and the other campaigners. It was not always so. In its earliest years,

United Fruit was seen to be all but heroic. Exploitation was the only game in town, totally in sync with the cultures of the big industrializing powers. Central America was the new Wild West. The isthmus between Mexico and South America was there for the taking. And apart from hurricanes, floods, droughts, disease, venomous snakes, and impenetrable jungles, substantial swaths of flat lowlands lay along the coasts with good soils and a dependably hot wet climate, ideal for growing bananas.

England, France, and Germany, along with the United States, saw the isthmus as the crucial passage to exploit resources and labor in the Pacific. In fact, ever since the Central American nations had gained independence from Spain, all the major powers on both sides of the Atlantic were driving hard for a share in the tropical bounty—and a faster trade route to the Pacific. As the twentieth century approached, and with it the earlier era of globalization we call the Industrial Revolution, the railway, the steamship, wireless communication, and refrigeration came together to put exploitation of natural wealth within reach of those with enough audacity to try their hand at empire building.

It was in this heady brew of opportunity and risk that entrepreneurs, buccaneers, and mercenaries from Europe, Asia, and the United States converged on the steamy coastal towns and smoldering volcanic mountains of Central America. In this period of intense exploitation, the concept that the tropical forest was a treasure to preserve for future generations would have been laughable; the tropics were a jungle, and the law of the jungle prevailed.

During the final two decades of the nineteenth century, Germans established enclaves in Guatemala and closed deals in Costa Rica and Honduras. The British too were everywhere throughout the region, setting up trade and investment outposts all over the isthmus and keeping a careful eye on their colonies in the Caribbean. American entrepreneurs were investing many dollars in gold mining, banana cultivation, and timber. France was taking the first steps in its grand plan to build a canal across Panama.

Even before these closing decades, a few men emerged independently as key players in the development of the Central American banana exporting business: Captain Lorenzo Dow Baker from Cape Cod; Minor Keith from Brooklyn; Andrew Preston from Boston; and from

Growing and shipping bananas in the early twentieth century.

New Orleans, Samuel Zemurray, an immigrant from Bessarabia, and the three Vaccaro brothers, originally from Sicily.

In 1870 Baker, a sea captain, arrived at Port Antonio, Jamaica, in his schooner. In a market there he noticed an exotic bent green fruit. An entrepreneur at heart, Baker saw opportunity. According to legend, he bought 160 unripe bunches, hoped for fair winds, and brought them back to New Jersey. There he sold the barely known fruit to eager merchants for a storied two dollars a bunch, giving Baker a tidy profit. The sea captain was about to become a banana baron.

At virtually the same time, Keith, a 23-year-old from Brooklyn, with his two brothers and workers from the Caribbean and from New Orleans jails, was helping his uncle—under contract to the Costa Rican government—to hack a railroad from the capital of San José to the Caribbean coast through an all but impenetrable tropical forest. Keith planted bananas on land along the tracks as ready nourishment for the workers. Unfortunately, as the tale is told, the railroad at first was unable to command an adequate number of paying customers to make a profit. Desperate for an alternative to the passengers, Keith thought of bananas. He risked hauling the fruit by rail to Limón, the port city, and then by ship to North America. The pieces that were to become United Fruit were falling into place. A decade or so later, Baker and Keith joined forces with Boston entrepreneur Preston to establish the Boston Fruit Company. And in 1899 Boston Fruit set up shop as the United Fruit Company.

That same year the Vaccaro brothers (Joseph, Luca, and Felix), along with a relative, Salvatore D'Antoni, lost their orange groves in Louisiana to an untimely freeze. They began to bring bananas into the States from Honduras. Relying on Italian immigrants who had set up farms in Honduras in the 1870s, the Vaccaros at that stage were small enough that United Fruit ignored them. They were able to stay in business because they dealt honestly with the small farmers. By 1906 the Vaccaros had incorporated their business, marketing tropical fruit, running a short-haul railroad, and bringing passengers by ship from Honduras to New Orleans, Baltimore, and Mobile. The company averaged something like one shipment of bananas a week, mainly to New Orleans—a total of a million stems a year.[2]

Not all European immigrants in the United States were so lucky. Only a year after the Vaccaros set up shop in Central America, the International Ladies Garment Workers Union was founded during fierce union agitation in New York City over the plight of immigrant women workers, many of them Jewish and Italian, in the sweatshops of the Lower East Side. The infamous Triangle Shirtwaist fire, in which scores of women jumped to their death or were trapped in a building whose fire doors had been locked by the owners to prevent theft, helped to consolidate the movement. It was a harbinger of the union movement that was to be a major factor in the plans of the American banana companies.

Musical Chairs

Samuel Zemurray earned the sobriquet Sam the Banana Man in Honduras. He underwrote a coup there in 1905 that brought generous concessions from the government for his developing banana business. The Banana Man pioneered new plantation techniques and in 1915 began production from large new landholdings in the Motagua Valley on the disputed border between Honduras and Guatemala. In 1930 United Fruit bought out Zemurray's holdings for $31.5 million; in 1933, as the largest shareholder, he became the company's managing director. By then the company owned plantations the size of Switzerland in Central America and the Caribbean. Zemurray's reign, a mixture of colonial benevolence and bold business dealings, had begun.

By 1924 the Vaccaros had expanded to a range of businesses from sugar mills to liquor-making companies, and had reorganized as the Standard Fruit and Steamship Company. In the 1960s Standard Fruit became a subsidiary of Castle & Cooke, a seventy-year-old Hawaii-based company with interests in pineapples, sugar, nuts, and real estate. Now doing business as the Dole Food Company, Standard Fruit is the only company that did not eventually succumb to the blandishments of United Fruit. Dole and Chiquita remain the largest U.S. marketers of bananas. They are competitors everywhere in the industrialized world, with the lead in total banana sales bouncing back and forth between them.

Fresh Del Monte Produce, Inc., is currently the third-largest marketer of bananas in the world, but numerous and complicated changes in actual ownership through the years have the feel of a slow-motion shell game. Established in 1892, the company got a foothold in the Central America banana business in 1958 as the result of an antitrust action filed against United Fruit in 1954 by the U.S. government. The final judgment demanded major divestment of assets by the company, which by the midfifties owned some 1.7 million acres in the American tropics. United Fruit was ordered to dispose of at least a third of its banana holdings to an independent company. El Pulpo elected to sell its total Guatemalan holdings to Del Monte.

The Del Monte brand has an intricate modern history. The first major conglomerate to buy Del Monte was R. J. Reynolds, which acquired it in 1979. A decade later, Reynolds separated the fresh produce business from the canned fruit and vegetable business and sold the division. The buyers were the British firm Polly Peck and Mexican investor Carlos Cabal Peniche, who jointly owned Del Monte Produce until he fled his country in 1994. (The government had accused him of making $700 million in fraudulent loans, essentially to himself.) Cabal, the high-flying head of Banco Union and Banca Cremi, disappeared in 1995 when the government took over his failing banks as part of a wider $100 billion bailout of the banking sector during Mexico's 1994–1995 financial crisis. After he fled, Polly Peck declared bankruptcy, at which point the Mexican government took over Fresh Del Monte. Until recently, Cabal was held in a maximum-security prison in Melbourne, Australia, fighting extradition to Mexico. Interpol had arrested him in a Melbourne suburb, where he had been hiding for four years. Ironically, on arriving in Mexico in 2001, he was allowed to walk free; a court had ordered him released on his own recognizance pending trial.[3]

In 1996 Fresh Del Monte was taken over by Abu-Ghazaleh, a Palestinian Arab with thirty years' experience in international trade, the last ten in Latin America. Abu-Ghazaleh's IAT Group in Chile paid Grupo Empresarial Agricola Mexicana $530 million for Fresh Del Monte. In the banana industry the company is still referred to as Del Monte. Some shareholders continue to contest the settlement.

The Heavy Gorilla

For most of the first five decades of the twentieth century, United Fruit was the 900-pound gorilla in the banana business. In 1899, after Minor

Keith cut the deal with Andrew Preston and Lorenzo Baker to form United Fruit, the company controlled 250,000 acres in Colombia, Cuba, Jamaica, the Dominican Republic, Honduras, Nicaragua, and Costa Rica—more than one hundred square miles of banana land. The company operated some three hundred boxcars and seventeen locomotives, covering a hundred miles of track between the plantations and storage buildings on the coasts. Even in this early period the company employed fifteen thousand workers.[4]

At the turn of the century, chartering or leasing ships from European owners was the standard practice for fruit companies. But United Fruit owned thirty-six steamers carrying bananas to southern ports such as New Orleans and Mobile, and to still more ports on the eastern seaboard. The company was a pioneer in refrigerator vessels. Its first was the SS *Venus,* launched in 1903. That year United Fruit imported more than a billion pounds of bananas from its Caribbean holdings.

In 1906, at a time in history when science was exploding into commerce, United Fruit set up two radio stations in Nicaragua and already operated a 200-foot transmission tower in Panama. Then in 1919 United Fruit teamed up with some companies whose names still resonate in the communications business (General Electric, Westinghouse, AT&T), with the aim of controlling the embryonic radio industry. These companies already owned the patents for radio; all they needed was a network to broadcast. They formed the Radio Corporation of America (RCA), which in turn launched the National Broadcasting Company (NBC).[5]

United Fruit, even in those early days, was reportedly different from its rowdy competitors and began its long mutual admiration with the United States government. For example, in 1914 the American envoy in Honduras wrote home that United Fruit was a "distinctively American concern" employing "clean-cut young men brought out from the States" and not the "American beachcombers employed largely by the other companies." By 1915 this distinctively American concern was working a fleet of almost a hundred ships transporting tens of thousands of bunches of bananas on each trip.

Between 1930 and 1940 the company paid $40 million in taxes to governments in Central America (accounting for about half the tax revenues of Costa Rica and Honduras) and an estimated $200 million in wages to banana workers. In a mixture of cultural paternalism, business necessity, bigotry, and outright power, United Fruit operated eighty-one schools for children of their workers but imported more tractable workers by the hundreds from Caribbean islands, while slapping down budding union initiatives and refusing nonwhites access to company hospitals. Critics still point to United Fruit's high-handed style of dealing with Central American governments. A quip credited to Zemurray—"a mule costs more than a deputy"—sums up El Pulpo's banana republic style. It was a style that in 1954 was to descend into infamy when the company cosponsored a coup of the Guatemalan government hand-in-hand with the communist-obsessed U.S. Central Intelligence Agency.

El Pulpo Goes Too Far

Chiquita's current management is generations beyond the seamy Guatemalan era of the fifties, but it is almost impossible to read a press account of Chiquita even now that does not refer to those shameful days. At the height of the anti-Chiquita campaign, Alistair Smith and the other adversaries of dollar-banana multinationals pummeled the big banana company with this fifty-year-old story. Beyond the irresistible bashing potential inherent in the Guatemala chapter, a number of the institutional actors of the 1950s continue to be crucial for Chiquita today. These include warring elements inside the Catholic Church, socialist forces in Europe, and the international trade union movement.

The Guatemala debacle started when Juan José Arévalo took office as president of Guatemala and immediately began to move the country toward democratization. In 1945 large foreign-controlled companies employed the vast majority of workers in Guatemala. Some forty thousand Guatemalans depended directly or indirectly on United Fruit and its various railroad and other subsidiaries. With so many workers, United Fruit was bound to be deeply affected by the new labor code installed by Arévalo in 1947. Under this crucial transformation, the government legally barred itself from routinely supporting large-farm owners and other employers. The code was modeled on the American Wagner Act (the National Labor Relations Act of 1935), cited by some as the single most important piece of U.S. labor legislation in the twentieth century. It was enacted to eliminate employers' interference with the autonomous organization of workers into unions, and to establish the federal government as the regulator and ultimate arbiter of labor relations. Urban workers were guaranteed the right to organize unions, to bargain collectively, and to strike. The code set minimum pay rates, regulated child and female labor, and established a mechanism to adjudicate disputes. These measures are also incorporated in the International Labour Organization's basic tenets and are increasingly demanded of multinationals operating worldwide.

Land reform was overdue in Guatemala, but political foes of Arévalo seized on the new code to accuse the government of communist leanings. At a time when the Korean War was fresh in American memories, and American soldiers had fought against communist rule in Asia, this threat got the attention of the Federal Bureau of Investigation. Files were started on the top figures in the Guatemalan government. Plantations of bananas (mainly United Fruit) and coffee (mainly owned by Guatemalan elites) constituted less than 0.005 percent of all the farms in the country, but contained at least half of the farmland.[6]

Though moderate and gradual in some ways, the Arévalo reforms set off political turmoil, including the death of one contender for the presidency, and almost continuous labor action against United Fruit during 1948 and 1949. When a substantial stash of arms was uncovered in a United Fruit railroad facility, Arévalo declared a state of emergency. Under pressure from all sides, however, he was forced in 1951 to relinquish control of the government to Jacobo Arbenz Guzmán, a military leader from the early days of the Guatemalan revolution.

Arbenz was no friend of the multinationals either. His dramatic extension of the reforms started by Arévalo was finally to provoke the infamous 1954 U.S.-backed coup, abetted by and waged on behalf of United Fruit, clearly the most powerful enterprise in the country. Far from a garden-variety dictator, Arbenz proposed to give Guatemala back to its people. He made it a priority to build a highway to the Atlantic that would end El Pulpo's monopoly on shipments in and out of Puerto Barrios. Moreover, Arbenz told senior United Fruit management that he would insist that the company accept his government as the absolute arbiter on labor and management disputes. He would also begin to explore ways to recompense Guatemala for United Fruit's role in degrading the country's precious land resources.

In a crucial action, in March 1953 Arbenz expropriated more than 200,000 acres of United Fruit company land (95 percent of which was at that time unused) near the Pacific Ocean. Although United Fruit insisted that it needed the huge tracts of fallow land as insurance against increasingly virulent plant diseases, critics then and now have countered that this margin was far in excess of the actual need. Arbenz offered compensation in the form of bonds worth about double what the company had paid for the land, and a similar amount for two additional tracts on the Atlantic side—almost 400,000 acres altogether.

While the Guatemalan government and the U.S. State Department squabbled over details of this deal, United Fruit was calling in some big chips in Washington. Its action unleashed a formidable who's who of movers and shakers, as well as a massive public relations campaign engineered by Edward Bernays, the heavily networked publicity genius.[7] The goal? To persuade key people in the American power structure that the Arbenz government needed to be overthrown. U.S. interests needed to be defended from an imaginary tidal wave of post–World War II communism that threatened the fundamental values of the United States—and, in the bargain, the vast economic interests of United Fruit. The company rolled out a legion of arm-twisters and public relations troops at a cost of over half a million dollars annually, to sell the idea that a monstrous enemy was abroad in Guatemala.

The payoff was the muffling of any congressional sympathy for Guatemala. A sequence of surreptitious Keystone Kop adventures resulted in the equivalent of an invasion from quasi-mercenary troops headed by a puppet Guatemalan exile, Colonel Carlos Castillo Armas. Familiar figures such as Howard Hunt (then the CIA chief in Mexico) were involved, and these actions were quietly approved by secretary of state John Foster Dulles and not so quietly directed by his brother, CIA director Allen Dulles. United Fruit provided ships and other logistical support. On June 27, 1954, Arbenz was deposed, Armas was installed, communism was held at bay in Guatemala, and a sorry day for America and United Fruit entered the history books.

The CIA later got a new lease on life, emboldened to gear up for the Bay of Pigs debacle in Cuba, but the fruit company did not fare so well.[8] Despite major arm-twisting, the company could not convince the U.S. Department of Justice to drop a major antitrust suit threatening operations in Guatemala. (Some believe that Justice was pressured into this suit to allay worries that the State Department and United Fruit could be seen as being in bed together.)

In 1958 United Fruit accepted a consent decree that forced it to surrender some of its local business to Guatemalan interests and acres of its land to local businessmen. Another suit forced it to surrender ownership interest in the IRCA Railroad Company. Finally, in 1972, the company sold almost all of its Guatemalan interests to the Del Monte Corporation. This transaction gave Del Monte a solid foothold in the rivalry with Dole and Chiquita that continues to this day.

Although 1954 did not mark the end of U.S. interference in Central American politics, it was clearly a turning point for El Pulpo. Ahead lay many twists and turns, especially the big build-up in production on Carl Lindner's watch and the pioneering accord with the Rainforest Alliance in 1992. But today's Chiquita, much more than its rivals, has the Herculean task of getting—and keeping—that monkey of history off its back.

5

Strange Bedfellows

G reen labeling had been around a while before Dan Katz and his colleagues at the Rainforest Alliance determined to attempt their approach to tropical forest management. Green labels had been tried almost entirely on products such as light bulbs and toilet paper. The idea that certifiers could go out into the forest or onto farms to evaluate and approve with a seal the actual day-to-day performance of private companies was really unproven when Katz and his associates tested it in the late 1980s. At about the same time, in response to pressure from citizens, local municipal councils in Germany banned the use of tropical woods for public building projects.[1]

If the Alliance's certification approach were to have any real teeth, Chiquita's farming methods would have to reflect the strict water and soil conservation and worker health and safety standards that its principles called for. By 1999 the certification script ran to several pages with specific targets and had been applied at Chiquita farms throughout Central America. We knew the going had been slow when Chiquita piloted the program in 1992, but at the time of our initial observations major changes should have been evident.

The first field investigation we undertook was on the heels of the four events of great moment to Chiquita: in February 1998 the Panamanian banana workers' walkout; on May 3, the stinging exposé of Chiquita's business practices in the *Cincinnati Enquirer;* later that same month, the conference in Brussels[2] that stiffened the resolve of the European governments to continue favoring their former colonies; then in the late fall of the same year, Hurricane Mitch.

According to the *Enquirer* reporters, among a long list of charges,

Chiquita had used toxic chemicals that were banned or restricted in the United States and in Europe; had brazenly crop dusted above workers and their families; and had otherwise attempted to bully its way around Latin America. Chiquita has never denied that practices on its farms and with the unions had over the years been as bad as—perhaps even worse than—any in the industry. True and credible reform would doubtless take years and cost millions.

These matters were on our minds as we threaded our way through the barrios of San José, Costa Rica's capital city. Engulfed in a haze of diesel fumes, growling trucks, and sputtering mopeds, we were passengers of Dave McLaughlin, Chiquita's senior environmental manager, who was at the wheel of a blue Toyota 4Runner. With full knowledge of our professional backgrounds, Chiquita's management had okayed the time it would take for the stocky twenty-year company veteran to accompany us on our investigation of its farms in Sarapiquí, Guápiles, in the humid northeast of Costa Rica.

Fifteen years earlier, Chiquita had undertaken larger-scale, controversial expansion of banana plantations in an area that had originally been tropical forest but over generations had seen a series of changes. Some early banana plantations had been abandoned, then cleared for cattle ranching, and had then been replanted to bananas to meet the huge demand anticipated from the European Common Market, and the opening of Eastern Europe and Russia.

As we left the jammed streets of San José and nosed up the fairly new highway over the Continental Divide and through the miles of forest-covered mountains, dormant volcanoes, and steep canyons of Braulio Carrillo National Park, McLaughlin answered our questions with candor and good humor. On the way up the pass we noticed a motley convoy of trucks loaded with thick logs making their way south. "I don't know for sure where those logs are coming from, but timber that size is mostly in the preserves. They're supposed to have permits, but I have my doubts," McLaughlin observed. The Costa Rican government (often more progressive than other nations in the isthmus) has severe rules against logging in the parks and preserves, but enforcement is uneven and, some say, shady dealings take place between certain government officials and the woodcutters.

McLaughlin, an American, had the credentials for the big job Chiquita had assigned him. He grew up in Colombia and Bolivia and had a broad political science academic background. In addition, he had a number of years of on-the-ground agricultural management experience in the tropics. His good-natured presence belied what we would come to recognize as a gritty determination going beyond a technocrat's take on the challenges a banana manager faces.

It verged on the incredible that his giant company, freighted with the old United Fruit baggage, could have broken from the pack to risk allowing the Rainforest Alliance—a comparatively tiny and outspoken defender of the tropical forest—to scrutinize its operations in every country where it had farms. In our professional experience, no other multinational in any industry had been willing to make such a bargain with a possible devil. The pact represented a highly unusual level of trust between an advocacy group and a multinational company.

We knew, by the time of our visit, that every farm in Costa Rica had been certified to Better Banana standards. In keeping with the principle of continuous improvement under the agreement, many by this time had been recertified a number of times. McLaughlin observed that the "profile" of pesticide usage had changed significantly. Now the company used only 10 to 15 percent of the most toxic chemicals (the U.S. Environmental Protection Agency's Class 1 and 2); the rest were much less toxic. McLaughlin later shared with us company statistical data confirming these reductions. He said that the proportion might vary over time as different pest explosions demanded response. The most toxic chemicals are utilized mainly to deal with nematode populations that attack the roots of the banana plant. On its own, Chiquita had eliminated the use of Bravo (chlorothalonil), a fungicide registered by EPA for use on a wide variety of crops. (The chemical continued, in 2003, to be used on banana farms owned by Chiquita's rivals and independent farmers.)

"With Better Banana we found that what we thought was impossible was possible. Getting rid of herbicides, including Paraquat—five years ago we would have thought that was basically borderline absurd," McLaughlin admitted. "Now we find that without the chemicals we're getting unbelievably strong root growth." Paraquat is a herbicide widely used for broadleaf weed control. It is a cheap, quick-acting, nonselective compound that destroys green plant tissue on contact and by movement

Dave McLaughlin, senior director, environmental and regulatory affairs, Chiquita Brands International, Inc.

within the plant. Employed for killing marijuana plants in the United States and in Mexico, it is also used as a crop desiccant and defoliant, and as an aquatic herbicide. As the Rainforest Alliance's Chris Wille observed, it's used "like water" throughout Latin America and elsewhere.

McLaughlin said, "Bravo and Paraquat are still fairly common among some of our competitors. Paraquat use is not allowed by the BBP, since it is one of the 'Dirty Dozen' most toxic chemicals. But Chiquita already banned this product in 1991—even before we started with the certification efforts. Bravo is still allowed under the Alliance's guidelines. It's an effective fungicide, but we feel that given its highly toxic nature to aquatic life and the fact that it causes skin and eye irritation, it is not a product we should be using. Chiquita has not used Bravo commercially since the late eighties."[3] (For chemicals banned under BBP, see Appendix B.)

If Better Banana certification was such a good idea, why, we wondered, hadn't Dole and Del Monte jumped on board? Dave believed that Dole never wanted to engage the conservation community. There may have been some fears that it would mean the banning of Paraquat. Dave opined that Dole officials didn't want to have an environmental group dictating what they could and could not use, and perhaps thought that the Rainforest Alliance would force them to label their fruit with a Better Banana sticker. "I think they misunderstood the issue completely," he said. "Dole's approach practically guaranteed a relationship based on mistrust. They could never take the certification idea at its face value."

In the early days, McLaughlin commented, the industry response to the local NGO firestorm was to form an organization called ECOBANA— an offshoot of the government/industry banana association called CORBANA. The new group did not have the credibility it needed, or the money. U.S. experts became involved and recommended huge expenditures on wildlife and toxicology. CORBANA began auditing all the farms in Costa Rica. It continued for a few years but without intensive or high-profile effort. Chiquita, which had broken from the pack and was working directly with the Rainforest Alliance, did join the CORBANA initiative, "but our primary focus was to get Better Banana up and running." According to McLaughlin, it was Chiquita that persisted in the idea of reaching out to the NGOs. Del Monte lagged far behind, as did the scores of independent producers. "Now," he said, "you've got some that are just top-notch. You'll see them tomorrow. And you've got some producers that are very, very bad . . . couldn't care less about environmental and social issues. Overall, I would say that the industry has made big leaps forward."

Our conversation was difficult to sustain, with stunning mountain landscape on every side as we climbed the side of Vulcán Barva. Emerging from the fairly daunting unlighted Zurqui Tunnel, the road spiraled down through the cloud forest of the park. McLaughlin handled the twisting road with composure, as macho drivers fought for the lead in the stream of traffic on the two-lane road. There were long moments of silence as the awesome spectacle of nature warred with the intense testosterone on the road.

Leaving the exciting mountain ride behind us, we cruised through gentler land of sugarcane fields and cattle fincas. "I've been encouraging the Rainforest Alliance to look into certifying these cattle ranches." Pointing to our right, Dave observed: "There's an example of the cattle pasture problem right there. That was all forest before. It's not managed. They don't manage, they don't maintain, they don't get rid of competing grasses. They just put up some fences and let the cows eat whatever's there." This was a tribute to the esteem in which McLaughlin held certification as a tool. We learned later that the Alliance had resisted approaching the cattle ranchers mainly because of not wanting to spread itself too thin early in the evolution of its program. Beyond bananas, it chose instead to focus on coffee, cacao, and citrus plantations to save trees and protect remnants of forests.

Bananas occupy less than 1 percent of Costa Rica's landmass, and provide jobs for more than forty thousand workers when all companies are combined, but fully a third of Costa Rica has been deforested for cattle. These operations produce very few jobs and add relatively little to the government coffers. As with bananas, McLaughlin felt that if American beef buyers such as McDonald's were to certify pasture maintenance there would be no need to put a label directly on the product. "They should just assure that their beef is being sourced correctly. If I were the hamburger companies, I would be worried about it. It's one of those issues that could explode on you," McLaughlin says, no doubt with his own experience in mind. At this point, the papers were not yet full of the farmer-hero José Bové and his jousting with McDonald's in France.

Some two and a half hours from San José, we reached banana country. Acre after acre of the huge-leafed herbaceous plants line the roads, extensive monocultures stretching into the distance. It is rarely clear who owns these farms, but there are hints. Distinctive chemical-laden blue bags enclose rows of banana bunches on some properties, but these are interspersed with white bags on most of the Chiquita farms, indicating they are chemical free. Otherwise, Chiquita blends into Dole blends into Del Monte, and in between are scattered farms of independents that sell to the big three. All the companies used the conspicuous blue bags and, until certification began, had routinely dumped them on the ground. Often drifting into the estuaries, tangling up fish and turtles

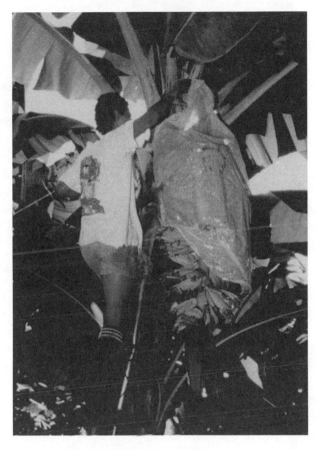

Bagging a banana bunch.

along the way, these conspicuous blue bags became symbols of banana baron intransigence.

If chemical or other polluted loads were found flowing downriver from these farms, it would be virtually impossible to tell which owner was responsible. Thus, if an independent producer were sloppy with his chemicals or his water management, any of the big three companies could be blamed—and vice versa, of course.

We left the highway and drove along a dirt road next to one of the

Chiquita farms. "This is another piece of forest that we have preserved intact all the way through. This area ran all along the river. There was evidence of water damage beyond the banks. We lost a lot of banana plants here because of the flooding," Dave remarked. "Most of this was cattle land, not forest, before we came in." He didn't deny that there *were* plantations that had taken the place of tropical forests. (In the fifties, United Fruit returned to old plantations in the Sarapiquí lowlands that it had once abandoned, gradually changing operations in the Pacific side of the country from bananas to palm oil. Some of the new plantations came from clearing forestland, but most were former banana farms.)[4] We came to a dirt road between rows of banana trees closed off by a railroad crossing–type barrier. After a few words in rapid Spanish between McLaughlin and the Costa Rican guard, we drove into a Chiquita farm.

"Bananas have to ripen in a certain way"

The banana plant as noted is not a tree. Banana and plantain are in the family Musaceae, genus *Musa*. The planting of bananas is timed to produce continuous fruit. The plants are genetically sterile and propagate asexually by putting out shoots, and these separate to form new plants. The dark spots in the center of the banana were once seeds. The stems continue to produce new generations of banana-bearing plants over many years. Three to four weeks after a root has been planted, the first shoots emerge from the soil. The plant grows a few inches a day until, in about nine months, it reaches an impressive full height of 15 to 30 feet, depending on the variety, with a leafy crown more than 20 feet in diameter. The stem is about a foot thick and consists of densely packed vegetative sheaths. The flower shoot with its many overlapping bracts begins to rise from the stem.

At first the small bananas grow downward. After a few days the bracts gradually fall away. Row after row, the bunches of bananas develop on the stem. Then, with exposure to light, the individual fruits—thanks to growth hormones—defy gravity and start bending upward, taking on their characteristic curved shape. Every bunch ready for harvesting can contain as many as seven to ten "hands," each of which has up to twelve or fourteen "fingers" (bananas). An average stem yields about

Bananas on the cableway. Harvesters put little cushions between the "hands" to keep them from rubbing against each other and scarring during transit.

150 bananas and weighs between 80 and 100 pounds. The fruit needs to be harvested green.

Plantations look more or less alike, with dirt roads, ditches, pathways among the plants, and above-ground cableways to carry the banana bunches to the packing stations. The undergrowth around each plant is regularly cleared, and mulch of dried banana leaves and stalks is added to keep down weed growth In some cases there are clear trade-offs in environmental gains (weeding by hand versus chemical control), but reductions in health threats from toxic chemicals are certainly in the worker's interest.

Banana workers, in pairs, harvest the heavy bunches, which are hooked onto cableways that carry them to the packing station. Easier on the workers than the former method of toting the heavy bunches on their shoulders, this procedure also protects the sensitive fruit from bumps and bruises. It is still hard work, especially harvesting in the tropical heat and mud during the rainy season. The workers pull about twenty bunches along the cableways above their heads—not a trivial load in the dank air. Once harvested, the leaf stems are cut down and used for mulch for the next plants. The so-called daughter and granddaughter shoots have already sprouted from the root of the mature plant. New shoots keep producing over several decades. Because the farms are set up so that the plants mature in a sequential pattern, the fruit can be picked all year round. Every season is banana season. Thus, the farms are producing even when bananas must compete with local fruits during the warm growing season in the north; they are ready to harvest whether or not there is, at any given time, a profitable market for them.

When the bunches are cut, starch biochemically converts to sugar in the pulp. Bananas contain about 20 percent starch and 1 percent sugar at the time they are harvested. In fully ripened bananas, the proportions are exactly the opposite. A banana is at full flavor when the skin has turned yellow with brown flecks. For the banana companies, the race against time starts at the plantation, when the green bananas are harvested.

Packing Station Action

Until the mid-1960s, bananas were shipped as whole bunches. Now a packing station is a feature of every plantation. As soon as the cable way gets the harvested bunches to the packing station, the bananas must be quickly graded for shape, thickness, and length, most often by teams of women. (The workers are supposed to wear rubber gloves to protect their hands from fungi, but not all of them follow this protective measure in the torrid heat.) Bananas that do not measure up—the culls—are rejected. They are sold on the local market, processed for banana puree, or used as animal feed. Companies have been known to destroy bananas on the farm to "adjust" supply and demand differentials. Better Banana guidelines encourage composting of banana culls, and we observed this on a number of Chiquita farms.

After thorough preliminary checks on arrival at the packing station, the individual hands are separated from the bunches and placed in a bath of water. Since the bunches contain much latex (natural rubber), cold water is used to stop them from "bleeding." Still, as much latex as possible is retained in the peel since it is essential for even ripening of the fruit. Any latex oozing from the peel can cause dark spots, and it stains hands and clothing; it needs to be washed off immediately. A great deal of water is used in the packing stations for this purpose, and conservation is a crucial issue. It can take as many as 15 kilos of water to wash a single kilo of bananas. The waste water containing latex and other organic matter is piped away. Pushed by the certification process, the company has tried filtration and recirculation technologies such as ozone treatment, ionization, and physical filters. In the absence of a workable technology, farms can still receive certification but must demonstrate at least some progress in addressing the problem. (In one low-tech experiment, water in the washing tanks was reduced without affecting fruit quality by filling the bottoms of the tanks with cement.)

A related issue is the common use of liquid chlorine in the wash water to counter possible bacterial contamination. Chlorine also removes the iron and manganese that can cause black spots on the skins. Widespread as its use is, chlorine must be handled with great care to protect the health of the workers, and it can damage the packing station infrastructure. Chiquita has been considering the use of substitutes, including hydrogen peroxide.

Next, the workers split the banana hands into smaller groups of fingers (five to eight bananas). The bananas are sorted for quality, washed, and then passed through a disinfectant shower. Then workers weigh the fruit, affix the company label onto each group of fingers, and pack them, green, in boxes for shipment. Inevitably, the background is one of loud popular music. From the packing station, the boxes are trucked straight to the harbor. When we were in Costa Rica, Chiquita was harvesting and packing enough bananas from these farms to fill three refrigerator vessels a week.

Men and women at the stations are paid based on banana volume packed over time, so the pace is brisk without being frantic. Workers are paid group rates and can work in teams of either two or three. To this extent the pace is under their control, not under the baton of a straw boss. The team method reduces pressure on the workers, and, as McLaughlin says, is better for the bananas. In the packing stations

Women banana workers selecting bunches for packing.

we inspected, the workers broke for lunch in a nearby shed at some previously agreed signal and returned the same way. We heard no horns or whistles or shouted instructions from management.

The ripening process starts as soon as the bananas are severed from the plant. To ensure optimum quality and freshness when the bananas reach the consumer, speed is essential. Within thirty-six hours at most, the exportable fruit, packed in the distinctive banana boxes, is loaded onto the refrigerated vessels. In the holds of modern freighters, the bananas are first put to "sleep" to interrupt the ripening process. The cold-storage chambers are maintained at a temperature of exactly 13.2 degrees Celsius (56 degrees Fahrenheit). If this temperature rises even slightly, the bananas start ripening inexorably. Once landed, the fruit is aroused from its slumber with a carefully calibrated amount of ethylene and shipped to stores according to a schedule that permits it to arrive at just the perfect state of ripeness for the shopper. Many wholesalers maintain ripening facilities themselves. To some extent at least, Chiquita can fine-tune the ripeness to accommodate supermarket preferences.

Chiquita's Farms

On one farm, Dave led us between the rows of banana plants twice our height, where we walked on a path of paving blocks made of re-cycled plastic bags—a Chiquita innovation. Each stem, drooping with the weight of the bananas, was kept erect against possible blowdowns with lengths of thin twine anchored to the ground. A powerfully built "pruner," stripped to his waist, passed by with a machete. The bananas were almost ready to be harvested. There was no cacophony of bird song, but an occasional bird could be heard flitting and calling. Silence and muggy heat were the realities.

McLaughlin pointed to the close-cropped weeds around the plants. "Herbicides were the main way we used to keep the weeds down. Now most of it's done by hand. It's more expensive, but it's part of the cer-tification standard." Here was a case where a trade-off between effi-ciency and environmental protection resulted in perhaps more physical labor but (for a worker paid hourly) in more money, too. The farm manager approached McLaughlin. After a private, intense conversation, the manager had a short discussion with someone on his walkie-talkie. "He said that there are unauthorized people somewhere in here," Dave explained to us. "We have a lot of valuable equipment, so we have to keep our eyes open." It crossed our minds that the unauthorized people might have been union organizers.

We had arrived at a time when the entire company was working through a major downsizing, a reaction to the considerable financial burden inflicted on the company largely by the stubborn trade impasse with Europe. Some two hundred employees stateside had been let go, and more and more responsibility was being delegated to the divisions and farms. "Every division is in a different stage of development," Dave noted, "being pushed literally from above as part of a whole empower-ment participation program. We're getting rid of a number of layers. Now the farm manager can make many of his own decisions, and there is a profit-sharing piece to it. We want to try to make people accountable and responsible but at the same time positive—getting them to participate."

We drove to another farm, where we were introduced to Nick, who had been a farm manager for just over a year. A friendly and relaxed young man, Nick used to work on shipping logistics but had for some time wanted to run a farm. With the additional autonomy of the managers, Better Banana certification had to find its way in concert with the downsizing. McLaughlin acknowledged a certain tension between the pressures to downsize and the environmental and worker safety template called for by the strictures of Better Banana certification. But the changes (in infrastructure, for example) necessitated by BBP could increase productivity, too. This could and did become a trade union issue. The targets were actually multiple: Nick and the other managers had to meet a baseline on productivity, quality and financial as well as environmental and social. "He can go beyond the baseline, and if he does it will show up in his evaluations," McLaughlin stressed.

"Certification is much more than a tech fix. It is ongoing—day in and day out. For instance, when we started off in Guatemala, we got the infrastructure in but we couldn't get the people part in gear at first. The people part is the hardest. It's not just an engineering job. You've got to have that commitment from the people. You can have the best infrastructure in the world but you can also not use it right. That's where you run into the challenge of changing the mind-set of the people on the ground." Dave said that to make certification work at the same time the corporate empowerment program was being deployed, he had had to spend hours trying to build understanding. "The willingness to execute usually follows," he observed.

Nick can try some things himself, but managers are allowed to buy only chemicals from the list of EPA-registered products. If they want to run an experiment with some other product out there, they do so in conjunction with the technical team in San José. "So we're not just wildly screening snake oils. We get a lot of salesmen coming out, you know, trying this product, saying it's natural, it's organic, and everything else. I say: 'I don't care if it's holy water, it still has to be registered.' From our standpoint, when we look at a product it's either got a general exemption from EPA or a requirement of a tolerance or it's registered. The list is fairly narrow," he noted pointedly.

As we drove through the farm, Nick said: "This farm has 210 hectares. We just got through planting ficus on both sides of the road right to the end. Helps with all the dust getting on the bananas. Not a certification issue, but we're doing it anyway." Pointing beyond the ficus, McLaughlin said, "This is still some of the original forest here but," gesturing ahead, "these are replanted riverbank fringes. It *is* a certification requirement. It stabilizes the banks and filters the runoff from the farm."[5] To our professional eyes there was no doubt that the river banks had been stabilized and runoff was under better control.

At farms in Sarapiquí, Guápiles, and Siquirres we learned that Better Banana certification had turned out to be a much tougher, longer haul than anyone on the farms in Central America had anticipated. Certainly nobody in Cincinnati had guessed the money and commitment called for. The Alliance itself was surprised. Starting from scratch and fine-tuning certification guidelines had turned out to be a high mountain to climb. It was the first time for everything; there were no precedents. Chiquita and the Alliance experimented with models of, for instance, solid waste traps in the service-water streams coming from the packing stations. Infrastructure had to be rebuilt. Profound changes were introduced in management and farm practices, especially pertaining to agrichemicals. The certification system was slowly being developed, and standards and protocols were put in place. Stakeholder input was being solicited. One reason everything took so long, McLaughlin told us, was because the company had to deal with a number of off-farm issues, like collaborating with the chemical suppliers to develop recyclable bulk packs. And in order to recycle the plastic bags, they worked with another supplier to come up with the paving blocks that we had walked on earlier that day.

The Voice of the Turtle

Much of the early certification effort was devoted to water quality. The guidelines required that water coming into the plantations and water going out from them be of much higher quality than had been the case over previous decades. In this region, watershed conditions cause both groundwater and surface water to eventually migrate to the Caribbean coast and to what has become a major nesting area for sea turtles.

Sea turtles, in fact, had been forced abruptly on the minds of Chiquita executives in Cincinnati. Protest letters had streamed in from all over the United States as a result of the campaign the Rainforest Alliance had engineered through the National Wildlife Federation's magazine, *Ranger Rick,* in February 1994. The campaign should have helped convince critics that the Rainforest Alliance and Chiquita were not exactly friends. But both the company and the Alliance turned out to be vulnerable to charges that the indispensable independence of the certification organization from the company was not so clear-cut after all. Critics, for example, charged that Chiquita paid the Alliance for the certifications, making an objective assessment impossible. Chris Wille says that while costs for travel and certifiers' time on the farms are reimbursed, substantial funding for the Better Banana Program comes from private foundations, mostly in the United States and raised independently by the Rainforest Alliance. Nevertheless, critics have persisted for years. (In our experience, even when it comes to government regulation, co-optation of the enforcer is a possibility that calls for constant vigilance.)

When we asked McLaughlin about this, he countered: "A lot of the Europeans like to assume that it's less than an arm's length arrangement, just to prove their point; but in fact, there have been issues between the Alliance and the farms, and we have to respect the Alliance's viewpoint. And it generally prevails." Alliance policy insists that the certifications be based on principles hammered out between the industry, scientists, and advocates, so that everyone is working from the same page. Certifiers from the Alliance can and do drop in on the farms uninvited, and violations have been discovered—and corrected. The policy is not to decertify any farm except as a truly last resort, but instead to work with the manager to improve his performance. McLaughlin commented: "We haven't had any big surprises that have thrown us for a loop. Generally, if they want to change something, we'll talk about it and try to understand it. When they changed the format in their manual of principles a couple of years ago, they tested it on us before the final version to see if it was doable or not."

In fact, the Alliance's timber certification program did decertify a big mahogany operation in Indonesia, and in 2002, after a surprise visit, Better Banana auditors warned several banana farms in Honduras that decertification was a definite possibility. Wille argues that the carrot is much better than the stick. The objective is to keep the farmer engaged in the process, and thus training becomes paramount. While this is true of all farms, it is crucial for the small farmer. The mission of BBP is sustainable development, not punishment.

As an example of a big change, McLaughlin told us about the only case where certification was unquestionably the cause of a labor conflict. In Bocas, Panama, workers formerly hand dipped the banana clusters in a postharvest mix, and the workers doing this task evidently refused to wear gloves. In order to become BBP certified, the farms needed to utilize mechanical application. To install the requisite spraying chambers, Chiquita had to remodel all the packing stations because the existing configuration did not allow adequate space for the chambers. This change required an additional investment of $2.5 million. But the workers preferred the extra hours gained from hand dipping, and the union went on strike over this issue for about eleven days. Finally the government stepped in and forced the union to accept the change because the goal was to protect the workers' health. We confirmed that these expensive spraying chambers had, in fact, been installed in the packing stations on the Chiquita farms we inspected in Costa Rica.

The Government's Role

From the outset, the Costa Rican government was supportive of certification, but it has never played a significant role in ensuring the integrity of the BBP certification. If we ignore for now the obvious pressures against enforcement arising from the government's long dependence on the banana industry, in reality there are no regular public reports comparable to those of the EPA to measure factors such as water quality. While its watchdog function is minimal in Costa Rica (but reputedly stronger than in many other Central American countries), the government does play an ongoing role. For example, the Ministry of Health regularly conducts sanitary inspections of all farm kitchens and mess halls. Inspectors also review conditions in the so-called bachelor quarters. Among other reviews they monitor preventive control for dengue fever and malaria. The government also provides recommendations on the use of safety gear. If asked by the banana companies, the government will test for water quality, but the test results are not made public. Penalties for violations are rare. To the extent that certification references national laws, it becomes a de facto enforcement measure.

The government is involved in a number of other ways, and in the early days when certification was first talked about in Costa Rica, it was thought that the government might play a rather active role. McLaughlin tells a story that probably depicts the more usual relationship between the government and the company. "When we first started certification, I remember the ministry's asking us for our designs on things like banana waste trenches, not because they were questioning them, but because they wanted to copy them. To some degree I think we provided them with a road map for what should be done to reduce impacts. Other than getting permits, we really were never ordered to install infrastructure—except for one meter of tiles on the urinals." (On the farms we did identify a number of these waste trenches.) He said that if we asked the government for a list of violations, we might learn that one or another banana company was cited for deforestation without permits. Or that a large independent might have employed the time-honored dodge of using contractors to avoid paying employee benefits. Or that some farms were experiencing disputes with the workers for violation of the labor laws. Shortly after our first visit, a coffee plantation owner in the province of Cortago was sentenced to five years in jail and fined $4,570 for dumping toxics into a river. The case made headlines since it was so unusual: under Costa Rican law environmental damages are not considered "serious."

Visible Changes

Water testing is a part of the certification process, and certainly certifiers and managers have held private conversations. Since the banana business is so brutally competitive, Better Banana certifiers do not make public technical aspects of company performance that might reveal trade secrets. Even so, certification has some very visible results, and a tremendous amount of information can be gained from alert and informed scrutiny (as most environmental auditors will attest). Such was the case with our observations.

Although the certification guidelines run to many pages, a well-managed banana farm from the point of view of environmental and worker health and safety boils down to some basics that are easy to call for and difficult to accomplish: conservation measures such as recycling water and preserving water quality; water testing and monitoring; replanting of native species; vegetating river banks for stabilization and filtration; creation or expansion of buffer zones between the banana farms and workers' houses, schools, and other facilities; management of solid waste; education and training of workers in chemical storage and handling; and the range of steps necessary to protect workers and the environment from toxic chemicals. On the farms we observed, we noted undeniable examples of all of these improvements in various stages of accomplishment.

What we saw, in fact, were generally well-managed, well-maintained storage areas for chemicals and tools; composting of organic waste; recyling of some (but not all) pallets; clean and relatively orderly packing stations; landfills that had replaced piles of inorganic waste; a program for pulping and reprocessing discarded bananas; and some replanting of forested corridors and stabilization of a number of riverbanks.[6]

We personally inspected clinics for the workers and new housing outside the farms which could be owned by the workers (a new program in cooperation with an international housing NGO). We also visited a modest childcare facility and saw separate showers and changing rooms, housing for teachers, video cassettes and monitors for training, and (in some places) neat hedges and gardens around the workers' houses. There were numerous soccer and baseball fields for the many "bachelor" workers, some (but not all) of whom end a hard day's work with another hard game under the lights. These benefits are part of a long history of companies providing social infrastructure—a vestige of the paternalistic heritage, increasingly burdensome for all the companies in hard times. And then Chiquita and the other companies—even if they feel compelled to assist with social infrastructure outside the farms—must contend with a jealous government that sees such changes as its prerogative. It is a catch-22: although the needs are clearly present, the government's capacity to satisfy them is severely limited by a variety of forces including its obligation to honor the structural adjustment measures insisted on by multilateral lending institutions.

Based on the Better Banana standards, one would expect to find, perhaps over a number of recertifications, many of the upgrades that we were able to document. The most visible improvements were in the handling of the various kinds of organic and inorganic waste—

for example, the plastic bags used to protect the banana stalks during growth; the plastic twine for helping to support the plants; the inevitable accumulation of culls and discarded plant material.

We inspected closed systems in company airports to keep toxics out of the ground and surface water. We found increasing use of vegetative ground cover for certain kinds of weed control instead of the steady application of chemicals. We observed replanting of native species in deforested areas to help restore biological diversity and new wildlife corridors created by strips of forest species.

A certified farm should pay strict attention to matters of worker safety and health. We saw without exception, for instance, careful isolation of storage spaces for toxic chemicals and control of access to these chemicals by lock and key. We noted showers to wash off chemical residues and laundries to clean protective gear. At the packing stations we observed decent bathroom and eating facilities. (It should be noted that the standards do not usually prescribe the "how," but only the desired end result. Farm managers have a number of options.)

Not as visible, but subject to public scrutiny, toxic chemical use would be expected to show a distinct downward trend over time, as would the presence of such chemicals in ground and surface water. Verifiable written documentation would go a long way toward confirming the trends. Since no substitute has yet been found to combat Black Sigatoka, the certified Chiquita farms we visited had mostly begun to increase the size of the buffer zones between houses, schools, and fields. The company had recently developed an A, B zoning system whereby workers were required to stay out of zone A during and thirty-six hours after aerial spraying, and similarly with zone B. We saw spray planes equipped with global positioning systems and antidrift nozzles that reduce fungicide usage and improve the accuracy of the spraying. We observed that most workers using toxic chemicals wore protective clothing and respiratory protection, but this gear is not comfortable in the heat and humidity. One government health official told us that she believed the single most important new contribution a company could make to worker health would be to develop protective clothing that is tolerable to wear in the tropics.

On a certified farm, continuous worker training in all of these areas had to be documented; while we saw some training video cassettes, we felt there could be improvements in this area. We did notice that some of the training materials included explanations of the ecological reasons for these precautions.

Of course, in 1999 the ILO-related workers' rights issues were not seen as a part of Better Banana certification. During our initial inspection many of these rights (freedom of association, equitable pay, prohibition of forced labor and child labor) were already under intense discussion between the companies and labor advocates. Chiquita's breakthrough agreement with the International Union of Food and Agriculture Workers (IUF), and with the Latin American regional banana workers' umbrella group COLSIBA, had not yet been signed, nor had the company's initiatives with the social certification group Social Accountability International been made public. As a result, before-and-after comparisons were not possible at the time.

Even at this early point, we did focus on the particular institution of Solidarity Associations, a uniquely Costa Rican answer to trade unions. On one of the Chiquita farms we talked with a senior worker who was a Solidarity official. He gave us a brief overview of the general workings of the labor association, but at the time we were more interested in the environment, health and safety conditions, and the Better Banana certification. The human rights revolution unfolding at Chiquita was still behind closed doors. We were, however, sensitive to the issues faced by women workers and their children, and concerned that failure on the part of both the Rainforest Alliance and Chiquita to address these and other rights issues would make them both continue to be vulnerable to critics. Even so, the singular arrangements between Chiquita and the Rainforest Alliance under BBP were already a giant step ahead of where Chiquita had been less than a decade before.

If we had visited noncertified Chiquita farms in 1992, or even up to 1996, in the other tropical countries where the company had operations, the picture would have been vastly different. As late as 1998, many independent farms and those owned by Chiquita rivals would have been environmentally and socially as problematic as they had been ever since bananas were first grown in vast numbers much earlier in the century.

On most of those farms we would have encountered chemical-laden, impoverished soil; big areas of deforested wasteland; gullies filled with plastic waste and discarded cord; rotten banana carcasses and stalks; shanties with outdoor latrines and no shower and changing rooms for workers to remove protective clothing; murky rivers and streams; heaps of discarded chemical drums and other containers; and a noticeable absence of plant and animal diversity.

While we may not have seen the sullen, angry faces of the workers so eloquently portrayed by former United Fruit employee Thomas Mc-Cann in his book *An American Company,* we would certainly have seen some of these very sorry labor conditions. Perhaps the scenes would not have been on a par with the worst stoop labor on California farms or in garment sweatshops in cities and slums around the world, but they were bad enough to warrant the alarmed cries of human rights activists. Some of the complacent companies simply failed to see the world changing before their eyes.

Not that we were witnesses to the equivalent of the chain gangs of the bygone American South. Although workers' housing and the domestic infrastructure on Chiquita's farms were in most instances adequate, there was often a stark contrast with life in the surrounding communities. Since housing on the plantations was assigned largely according to seniority, the remaining workers had to be bused to the farms from nearby towns, where living conditions left much to be desired.

On the farms, work went forward in a context of both urgency and care for the fruit, in which very special attention needed to be given to worker safety and health. Bananas had to flow from the farms in a steady stream that allowed for the handling and loading of a quarter of a million 40-pound boxes per vessel. As on assembly lines everywhere, the work could be both monotonous and dangerous. Some women complained that the bunches arrived at the packing station with a cargo of biting insects. Knives and machetes were everywhere. Clearly, no banana worker took one of these jobs in order to fulfill himself or herself, although many workers did make a career of their assignment. Without a decent salary and reasonable working conditions, it would be a dismal career indeed. Fortunately, we heard many reports—not all of them from Rainforest Alliance or Chiquita people—that most workers developed a certain pride in being on farms where the changes forced by certification were in their interest and beneficial to the environment.

We had a lot more to do on this trip. Most of the evidence to judge the value of the Better Banana Program would come from objective experts and responsible critics. It would also be vital to assess the integrity of the people, including the Alliance's Chris Wille, who had been on the ground from the beginning. Despite our obvious enthusiasm for the promise of certification, given our professional experience and the purpose of our mission, both McLaughlin and Wille understood that dissembling had a strong downside and that much could be gained by telling the truth.

The Other Side of the Bed

When we returned from banana land, we drove across San José to Moravia, the Alliance's outpost in an adjacent barrio, and sat down with Chris Wille, a tall, quiet-spoken Oregonian who looked younger than his fifty years. It was clear even in these early meetings that his mild manner masked a ferocious dedication to conservation and a conviction that certification by the Rainforest Alliance could make a real difference in the battle to protect remaining forest land and make sustainability a reality.

In 1989, when Wille settled in San José with his new wife, Diane Jukofsky, to publicize conservation efforts with the press in Latin America, they saw firsthand that at least some of the forest destruction was coming not from clearing for cattle, but from clearing for banana plantations. Chris was to become a central figure in the banana certification story.

He explained to us how the fledgling Alliance first engaged the banana giants. "By 1991, the expansion was really booming," he recalled. In the early nineties, the banana companies had bought up about a hundred square miles of Costa Rican grazing land and forests, in anticipation of new markets in Eastern Europe. Then in 1992 the Costa Rican Forest Service accused the banana company Geest of illegally cutting down primary and secondary forests in the Sarapiquí area.[7] Wille saw a challenging opportunity for certification and convinced Dan Katz that the fledgling organization could and should apply the principles it had successfully developed for forest management.

When unflappable Chris Wille first met with voluble Dave McLaughlin, they found a point where their different agendas could converge.

Chris Wille (right), *chief of agriculture for the Rainforest Alliance, speaks to a
prospective auditor in a training session.*

Their mutual ready humor, the built-in pragmatism of the certification
program ("So practical, it's radical" was the Rainforest Alliance's slogan
at the time), helped to bridge what could have been a giant gap in organizational cultures, motivation, and outlook.

We asked Chris to give us his version of Chiquita's epiphany. He told
us that all the big companies had sophisticated labs and test farms, as
well as bright people looking for ways to improve production, fight disease, and give themselves an agriculture-based competitive edge. "Almost no attention was ever given to environmental issues and very little
to social issues. Anything other than production simply did not come up
on the radar screen at that time. When we arrived in Costa Rica, the
word *environment* seldom crossed the lips of a banana farmer, and *en-
vironmentalist* was used by them only as a derogatory term."

He said some employees of the companies cared about "doing it
right" and about what we now call corporate responsibility. Some employees, mostly younger people who had not grown up in the banana

culture, were concerned about environmental impacts and sustainability. But, he said, "the industry was very concerned about pesticide issues, notably the big sterilization case" (referring to the episode that had happened some years past). In the 1970s, the industry had been coping with a highly publicized case involving dibromochloropropane (DBCP), an early chemical of choice to kill nematodes that attack the roots of the banana plant. This matter was to come to a head in 1997, when class-action lawsuits against Dow Chemical and the big banana companies—including Chiquita—were filed in U.S. courts by some 24,000 banana workers claiming that their sterility was linked to the chemicals.

Most farms had discontinued the use of DBCP in the seventies, but the companies continued to be blamed for agrichemical irresponsibility. The public began to have its doubts, too. Dow continues to dispute its liability, claiming users suffered because they didn't follow directions, but the company offered a $22 million out-of-court settlement to all the workers in the action. Because of lingering sterility cases, workers and perhaps some ambitious lawyers in Nicaragua and Costa Rica continue to seek redress. And the major companies continue to face the challenge of assuring safe practice with toxic chemicals by the hundreds of small independents from which they buy. Many activists still question how careful even the major companies are. Chemical salesmen continue to target the small farms, trying to sell anything they can. Reports have recently surfaced of Nicaraguan refugees using Paraquat to commit suicide. It is common knowledge that poor farmers the world over use such toxics to take their own lives.

"Even without the sterilization charges, the companies were facing a terrible image problem," Wille continued. "At that time they were still using other pesticides indiscriminately. In other words, spraying every week with everything, whether the farm needed it or not. Some environmentalists have always used that charge, but now it would be unfair," he added.

For its part, the Alliance was eager to prove the worth of certification in agriculture. All that was needed was its first client and a model appropriate for bananas. After certifying a small farm in Costa Rica and another one in Hawaii, Wille went after all three big players and the major independents, meeting with the Costa Rican representatives of the banana companies in Costa Rica in 1991.[8] The Alliance knew it could not tackle the job by itself. It knew that any number of well-meaning northern-based NGOs had retreated from earlier campaigns conducted with unthinking arrogance and without local connections and credibility. So Wille called on experts at the National University and invited all the main actors to the earliest sessions. In subsequent meetings, new banana standards were explored and ultimately developed. Some companies actually participated, along with Costa Rican NGOs, including one called Fundación Ambio that briefly partnered with the Alliance.

"The banana companies first learned about the certification program at the same time as everyone else," Wille said. At first, the Alliance was communicating equally well with all the companies, but "after Chiquita started adopting the program, the other companies stepped back." Dave McLaughlin and Carlos Vega, a Costa Rican national with an environmental management degree, were employees who really cared and were looking for ways to reduce impacts (and costs), but they found it difficult to talk about ecological issues in the banana culture. Chris continued:

> We gave them two important things: first, guidance—banana farmers didn't understand what environmentalists were complaining about because they didn't understand the impact their farms were having. We were able to outline the issues for them and then give them concrete, measurable, practical, and doable ways to fix, avoid, eliminate, or mitigate many of the problems. Of course, some banana company technicians helped design these new practices. Second, we gave them political cover. It was difficult for banana company people to talk to environmentalists. It made them look like wimps or apologists. But we gained enough credibility so that it was okay to work with us. That allowed the few closet environmentalists to "come out."

A further incentive for the companies was the local opposition. To the alarm of the Costa Rican government—now crucially dependent on the banana trade and on its carefully cultivated reputation as a green tourist destination—the expansion of plantations was causing local Costa Rican citizen groups to take up the cry against the Yankee Devil and his banana company capitalists. "Clearly, neither the companies nor the government could push this stuff under the rug any longer," Wille said. He summed it up this way:

The meetings we organized in 1991 were a big help. At first, the banana company reps sat on one side of the room and the enviros on the other, glaring across the table in disdain and mistrust. We used scientists—cool, neutral, speaking a language that both sides understood—to bridge the gap. A good thing about certification programs is that they allow all sides to begin talking at a common place, with shared objectives. The banana company reps were glad to be talking to greens who had ideas for solutions that didn't threaten, who really knew about banana farming and its economic, ecological, and technical challenges.

The trust with COBAL [Chiquita's division in the northeast banana country] grew by accretion. There were a thousand little moments and revelations and agreements and concessions. There was never a time when Dave and I raised a glass and said, "We've done it!" I think for Dave the rock was clearly at the top of the hill when Bob Kistinger saw the contrast between Costa Rican farms and farms in Panama, came back to Dave's office in Costa Rica, picked up the phone, called Cincinnati, and told the controller not to mess with Dave's budget.

Kistinger's instincts were borne out when both Chiquita and FDB, the Danish supermarket cooperative, retained independent experts to verify Better Banana's program by visiting the farms. First, the Danish environmental group Nepenthes, hired by FDB, endorsed Better Banana certification with minor reservations; then, Conservation International, working with Costa Rica's own independent scientific research organization, the Tropical Science Center (CCT), confirmed major on-the-ground progress. (Chiquita had hired Conservation International, which had hired the CCT to backstop the science.) More telling was acknowledgment by the sharply critical Foro Emaús, that through certification Chiquita was making slow but certain progress on the reduction of pesticides.

Enter Sustainability

Meanwhile, environmental protests were escalating in the United States. There were signs that international labor unions and a range of social and farmer advocates in Europe were raising questions about workers' rights and child labor on banana farms in Central America. These strands of protest came together with others on a global scale in 1992 at the United Nations Earth Summit meeting in Rio de Janeiro. Its aim was to meld environmental, social, and economic goals into an overall strategy for finding ways to achieve development globally without damaging the environment.

For the first time CEOs and other senior executives from many of the world's largest companies entered into dialogue with the thousands of nongovernmental organizations that also attended the United Nations conference. It was a golden opportunity for the governments to find a middle ground between the Third World's drive to climb out of poverty and the industrialized world's (belated) environmental concern.

The conference gave the world that polysyllabic rallying cry "Sustainable Development" as it sought to bring businesses, environmentalists, and all sectors of society together with governments in common cause. Although it was no more than a blip on the screen of the American public, major media all over the world picked up on the Earth Summit and probably added grist for local activists in San José. The new international mood no doubt also played a role in Chiquita's decision to reach out to the Rainforest Alliance. In Costa Rica it was not merely the local pressure, it was also the anti–dollar-banana voices raised in Europe that helped develop the momentum that eventually led Chiquita to certify. At Cincinnati headquarters, though, the mantra repeated over and over by Chiquita managers was that, above all, they wanted to "do the right thing." It was a slogan that, as researchers and publishers, we had been hearing for years from beleaguered company executives. (Note that the word "right" carries a helpful bit of ambiguity. It may sound like morality, but it could also be correct business practice.)

According to Wille, the process in Costa Rica was gradual: experimenting with the certification program, testing the standards, gauging the reaction of the environmental community, and watching costs. "Little successes" began to add up. Chiquita "never got the applause they deserved, but they knew they would have to make the changes someday, so why not now? Why not get a jump on the competition?" Cincinnati's support grew slowly, then boomed once senior executive Bob Kistinger became involved. When Kistinger "flipped," Better Banana was at last off and running. At this point, the European trade restrictions on dollar bananas were pulling Chiquita into an ever-deepening financial hole as the costs of major investments in the expansion struggled against drastically lower-than-expected earnings from the European market.

Nevertheless, Wille believes that Kistinger's enthusiasm led Chiquita to finally agree to expand the Better Banana certification to all of its owned farms, with an ambitious 1999 target. Business analysts and rival companies were astonished at this decision. Although the Alliance had outstanding and literally unique experience in certifying timber management, it really did not know much about bananas in the early days of the engagement. Fortunately, appropriate technical advice was available from the National University and the companies.

Wille recalls: "All the companies were using our standards as guidelines by the end of 1992, and CORBANA had lifted them to create its own program. The dance was a slow one. You don't win anyone's confidence overnight." He said that the situation was especially tough because the critics of the banana business—"greens, scientists, the Catholic Church, unions, and gadflies—had been slinging arrows at the producers, mainly the transnational companies, for years." The companies hardened their resistance: it was especially dangerous for banana farm managers to be seen with environmentalists. "Our field technicians met with company technicians quietly."

"Both CORBANA and the Rainforest Alliance learned a lot from the experimental tropical research institute EARTH," Wille says.[9] "We were the first movers, but EARTH was close on our heels in terms of timing and a fountain of good ideas." Although the CORBANA project was a way for the other companies to diminish Better Banana, it did have some positive value: it gave companies that shunned Better Banana "something to cling to." Months later Kistinger told us: "When we began to put out feelers in the conservation community, the Rainforest Alliance was the only organization offering a certification program specifically for bananas. That was a big plus. Also, remember that we had seen this coming. We knew we had a very big hill to climb. Our one-hundred-year history was never going to go away completely. We saw this as a marathon, not a sprint."

Wille reports that in the beginning, predictable tension arose from this experiment. The Alliance wanted to put a certification label on the fruit. Chiquita, understandably, with several decades of building the brand, refused to budge on this issue. The other significant conflict was the company's absolute refusal to discontinue the aerial spraying of fungicides against Black Sigatoka, the virulent leaf spot fungus. Like certain vineyard blights, the fungus can wipe out a farm in a matter of days. This problem has plagued the industry for years. Acquiescing was a major concession for the Alliance, but the alternative was to lose the chance to raise the bar on many other more tractable environmental problems, over thousands of acres of land. An incredible opportunity for agricultural certification would be lost. It was a hard but calculated risk for which the Alliance knew it would take considerable heat.

While the Alliance eventually had to yield on these points, it nonetheless hung tough on increasingly stringent guidelines—ultimately running to many pages—that called for very demanding changes on the ground. The company stuck with the Alliance through these early years even when it couldn't brag about its compliance. It had to hold its fire. Without 100 percent of its farms certified, the company could not guarantee that a Chiquita label stood for environmental responsibility. For credibility, if you certify one banana farm, sooner or later you must certify every banana farm.

Fortunately, done right, conservation can save money too. McLaughlin understood early on that there were measurable economic paybacks from the changes forced by certification. Recently he estimated that "in rough terms, we have seen huge reductions in herbicides (about 80 percent), nematode control has come down by about 50 percent, and postharvest fungicides by 50 percent. These reductions are worth over a million dollars annually." McLaughlin and Wille also point to the huge benefit to productivity when workers feel they can strive together toward an important goal.

Neither of these strange bedfellows, Dave McLaughlin and Chris Wille, could possibly have foreseen what enormous obstacles lay ahead. The daunting list included the complex and stubborn trade regime in Europe, a number of aggressive European activist groups who were to attack Chiquita (and therefore the Rainforest Alliance) for environmental double-dealing, hurricanes, currency devaluation, and fierce competition—some of it from companies with far fewer environmental and social scruples. Better Banana certification was not designed to cope with any of these. It was especially not designed to address trade union issues such as freedom of association, child labor, and other demands arising from an increasingly militant international labor and farmers' rights movement.

As Wille, the ardent conservationist, candidly put it early on: "We have reached the point where we can competently address conservation imperatives and some important worker safety and health issues; but when it comes to the emancipatory things like freedom of association, that's just not my bag." Wille has had to educate himself on these workers' rights issues. Both he and McLaughlin were also learning to cope with competing certification schemes like Fair Trade and organic labeling. Both, like Better Banana, were backed by a strong moral authority—though with a social, instead of an environmental, coloration. These labels were making modest inroads into stores in Europe, perhaps owing to their resonance with shoppers who felt their roots deep in the history of protest in Europe. Fair Trade in particular, with its built-in anti–free-trade message, was a force to contend with.

In fact, Fair Trade, with its connections through Banana Link to the Latin American banana workers' unions, may have been a factor in Chiquita's decision to reach out to COLSIBA. Just prior to our visit in March 1999, Chiquita met with COLSIBA for the first time. Perhaps it was in this meeting that both sides were able to find common ground. In any event, the possibilities for successful negotiation of the astounding pact between the international food workers' union IUF, COLSIBA, and Chiquita were literally being put in place as we toured Chiquita's farms. It would turn out that Chiquita's risky experience with Better Banana would herald a promising end to a hundred years of labor struggle in the banana fields of Latin America.

6

Grass-Roots Snapshots

It was inevitable that Chiquita's fortunes would sooner or later be caught in the escalation of antiglobalization protests surrounding the birth of the World Trade Organization and its crucial role in arbitrating world trade. As it has turned out, meshing environmental and labor standards with the considerations of the WTO—set up to referee trade relations among governments—is fraught with thorny conflict, since national aspirations toward environmental protection and labor rights can be most often construed as restraint of trade.

While Third World representatives were being excluded from the private negotiations among rich countries inside the halls, explosive protests on the city's streets in the late fall of 1999 virtually shut down the international conference of senior trade officials convened in Seattle by the fledgling WTO. In addition to the gas masks, billy clubs, tear-gas cartridges, bulletproof vests, and other urban riot gear now routine at global gatherings of movers and shakers, hilly downtown Seattle was swarming with activists of all ages and many persuasions. They carried provocative signs and placards with slogans such as "Resist Corporate Monocultures!" and "WTO is on the fast track to corporate world dominance!" A recycled wooden skiff pleaded "Save Mono Lake," an artifact of a monumental skirmish of bygone days. Throngs of resisters, most of them stubbornly well behaved, had donned costumes: turtles, ears of corn, bananas, and—with a grim sense of humor—a 12-foot-high condom, unrolled and reading "Practice Safe Trade." This antiglobalization fever has spread via the Internet throughout the world, fermenting in

The battle of Seattle, 1999.

seedbeds of unrest and sprouting in the streets of whatever world capital is prepared to risk the fate of the "Battle of Seattle." Extreme crowd control measures have muffled recent protests but sharpened protester anger.

One such seedbed was in Costa Rica in the late 1980s, at the time of the rapid banana expansion there. The banana country north of San José had a recent history of social and environmental activism that pierced the ranks of dollar-banana managers and at first caused them to circle the wagons defensively. Chris Wille and his colleagues would finally break those ranks and exploit the clear banana farm certification opportunity.

Voices of the People

In our Costa Rican investigations, we met with and interviewed a wide range of activists, government officials, scientists, and corporate managers. The following portraits set forth the environmental and social concerns of a number of the main players. In them we have tried to depict the flavor of the period and the crucial tension points.

Protests had been rough in Costa Rica during the late '80s and early '90s. The situation had calmed down enough by 1999 that we could sit on a steamy March afternoon in the town of Siquirres with one of the key figures to explore the roots of the unrest a decade before. A Catholic priest, Padre Gerardo Vargas, was a stocky man in his forties with a full black beard and no clerical collar. The padre had a smile that can only be described as beatific. But his purpose was undeniably resolute, and we knew he was taken very seriously by Chiquita and the Rainforest Alliance.

We met in a small adobe-style rectory behind the church that doubled as a classroom and a place of refuge for the poorer members of Vargas' flock. Dominating the small room was a huge primitive mural depicting the European conquest of Central America. With golden-sailed galleons in the background, a martyred hero reminiscent of Mexico's revolutionary hero Father Hidalgo leads campesinos out of bondage—hopefully toward a brighter day for the small farmer.

Padre Vargas is the leader of Foro Emaús, a network of small-farmer sympathizers, women's groups, trade unions, Catholic Church leaders,

Padre Vargas, leader of Foro Emaús, the Costa Rican protest group.

university teachers, and environmentalists, that is named for the town of Emaus in the former East Germany. Foro Emaús grew out of the 1980s unrest, and Vargas readily admitted to strong ties with a German social movement.

The padre proved a gracious and knowledgeable host. Over a cup of strong Costa Rican coffee, he patiently related the history and purpose of his group. We noted that he never singled out any banana company in his narration. He began with what was to become a familiar refrain: together with cattle raising, banana production is responsible for the destruction of forests and for other serious environmental damage. Added to this, he believed, was exploitation of the workers. This process was accelerated by the banana expansion decree, issued by the Luis Alberto Monge Alvarez administration in 1985, which transformed the production and commercialization of the fruit in Costa Rica.

Until the 1970s, Padre Vargas remarked, the president's office and the offices of the banana companies were in the same building. "National banana companies [the independent producers] are owned by people who have been ex-presidents, politicians, congressmen, people who move from government into the banana industry. The people who are selling the image of Costa Rica as a wonderful tropical paradise are the same people who have an interest in the banana business," he said.

The 1985 conversion from pasture, second growth, and original forest had not been a modest expansion. The plan had included government incentives for companies. Vargas stated that the overall area planted in bananas had grown from 25,000 hectares to 50,000 hectares by 1994. Banana production had doubled in less than seven years, up to 103 million boxes in that year. He told us that this expansion was not by the multinationals alone: in nine years local producers had increased the area of their farms to some 51 percent of the total. At this point, a large number of displaced Nicaraguans had arrived on the banana plantations, where, Vargas said, they were exploited even more dramatically than the Costa Rican workers.

These big changes in the northeast banana country had caused the Catholic Church of Limón to issue a pastoral letter on December 25, 1989, entitled "Uncontrolled Banana Expansion." The letter acknowledged the positive outcomes for the Atlantic zone: much increased employment at decent wages, an increase in commercial activities and services, and overall more resources to benefit the regional economy. It also cited a long list of negatives: alcoholism, drug abuse, prostitution, violence, "individualism," instability of family life, increased economic dependence on transnational corporations, and consolidation of land tenure in a few large companies, thereby squeezing out the small farmer. The letter also referred to the loss of workers' representation in legitimate trade unions, a rise of consumerism and the resultant loss of a sense of community, pesticide poisonings, forest degradation, and pollution of rivers and streams. Many of these issues would sooner or later be addressed by Better Banana certification.

The large-company–small-farmer issues are still highly relevant today. From the perspective of a small farmer, his world is encircled not only by banana plantations but also by various forces that can present themselves in policies of the government. These policies are dictated by "structural adjustment" restrictions laid down, for example, by the International Monetary Fund (IMF). They skew the economy toward government preferential treatment of export crops and export businesses at the expense of local farmers producing for local consumption. They also skew government toward short-term fixes to earn revenues in order to pay back international bank loans, and away from long-term investment in, for example, social infrastructure. This serious problem was central to the Seattle protests that connected IMF policies with the free-trade mind-set, a stance that dominates the thinking of controlling capitalist cultures to the north. All of this, of course, has not been lost on increasingly sophisticated small-farmer networks.

Vargas recalled that protests had been mounting over the years, and in June 1992 a large regional forum was convened with the participation of unionists, environmentalists, campesinos, indigenous peoples, and religious leaders. September 2 of that year, more than twenty-five hundred people marched through the streets of San José to protest the conditions on the banana plantations and in the towns of the Caribbean coast. The event brought confrontations with elected politicians and government officials. The posters, banners, flyers, and throngs of people made headlines for days. The demonstration helped validate the mission of Foro Emaús, although, Vargas admitted, the conservative Catholic diocese in San José, led by Padre Claudio Solano, took quite a different view of the matter—far more akin to the Roman Catholic resistance to social reform pressures in Europe.

The pastoral letter and the street demonstrations got the attention of the banana companies. Foro Emaús, as a vigorous proponent for workers' rights, continued its campaign right through the mid-1990s until certification results became palpable and Chiquita gained some credibility.

By 1999, the years of outreach to the activists by both Chiquita and the Alliance were paying off. Organized protests in Europe over market, union, and trade issues had become the dominant theme, though there were still pockets of anger in Costa Rica. In seeking out reliable and representative critics, we spoke with a long-time leader of a local umbrella conservation group. He operated a bed-and-breakfast in banana country and so was quite familiar with the issues. We sat with him at a table in his open-air dining room near a stream. A macaw in a cage nearby looked down at us with unblinking eyes. At first the owner was cautious and reluctant to let us tape the interview, saying teasingly that we could be CIA agents. Although he was joking, he evoked a potent code that harked back to the not-so-distant past when Central America was feverish with the U.S. war against communism and the dark days when the CIA engineered the overthrow of the Guatemalan government. While the communist residue in the culture is all but gone, one can still sense a distrust of the U.S. government's intentions, so bound up is U.S. foreign policy with a free-trade ideology biasing economic models toward export commodity crops.

A slender man in his middle forties, with black hair, a handsome mustache, and thick-rimmed glasses, our host cogently outlined the principal arguments against industrial agriculture and destruction of local farmland. His views on the environment would have resonated well with many of the bed-and-breakfast clients who used his establishment as headquarters for bird watching and ecotourist trips into the forest reserve nearby. They would perhaps be less sympathetic with his perspective on the consequences of free-trade policies for the lives and prospects of the poor farmers.

He fiercely articulated the plight of small farmers worldwide wherever big foreign agricultural operations penetrate into developing countries. When we asked him about the viability of the Better Banana Program, he scoffed, calling it greenwash and observing that Chiquita was paying off the Alliance. He did admit there had been what he described as small cosmetic improvements on nearby farms, but he then proceeded to weave a tapestry heavy with threads of real history, anecdote, and passion. He believed that all the big foreign companies care solely about profits and do not spend enough, for example, on chemical research. Their only motive is to control world markets and Costa Rica's future. He accused all the foreign companies—not only the banana companies—of destroying the tropical forest in his country to produce wooden pallets. The big multinationals are treating Costa Rica, like many other parts of the world, as a "throwaway planet." They have taken much of the wealth out of Costa Rica, he said, for the last 120 years. If it weren't for this, he said, Costa Rica by now would be a First World country. Among his specific charges was that Chiquita had blamed its abandonment of the Pacific Coast plantations on union strikes, but it had really left because the soils were saturated with copper-based fungicides that had been used in the past to battle banana diseases.[1]

Local subcontractors to the big companies, he charged, were getting rich exploiting Nicaraguan refugees. They paid the men for machete clearing by the hectare, but measured hectares as more than 15,000 square meters, as opposed to the correct 10,000 square meters. Both the government and the companies, he said, needed to stop deceiving the people. He noted that in 1992 the Costa Rican government had claimed that the U.S. Army Corps of Engineers was used to build some bridges and schools, but it was common knowledge that the real purpose behind their presence was to assist U.S. companies in the banana expansion.

Picking up another theme we were to hear often, the activist said that corruption existed at all levels of government in Costa Rica. He charged that politicians were getting something every time the government allowed an expansion. He also referred to past episodes of violence against conservationists. His sincerity was palpable; it was hard to doubt the depth of his conviction.

(In fact, conservationists can and do face violence in Central America. Perhaps this is less true in Costa Rica, but our guide and interpreter David Dudenhoefer, an American with considerable experience as a journalist in the isthmus, told us about an incident in Guatemala in which Arnoldo Ochoa Lopez and Julio Armando Vasquez Ramirez were killed in Puerto Barrios. Both had worked for CONAP, the Guatemalan equivalent of the National Park Service, and were involved in prosecuting illegal loggers. They had received death threats, and Arnoldo had barely survived an assassination attempt years earlier.)

The most important action the companies could take, said the innkeeper, would be to discontinue expansion of their banana farms so that the small farmer could begin to find a dignified place in the economy on the remaining productive land. His country is importing rice and beans, in fact, and many other agricultural crops. Bananas are being planted on the best lands. The next-best thing the companies could do would be to activate an aggressive program protecting the soil and the remaining forests. He said that the Costa Rican government should have started thinking about this problem fifty years ago. There had been absolutely no planning to preserve decent soil. In the past, the only way to acquire title to land was by clearing more than 50 percent of it. The government and the companies used to call that "improvement," but in reality it was the opposite.

As we were leaving, he offered us a couple of organic bananas picked

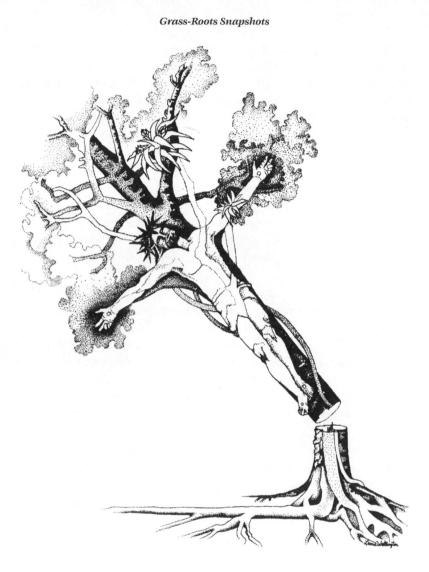

from a tree growing next to the dining room. He also gave us a white T-shirt depicting in stark black stenciling a Christ-like figure with a crown of thorns impaled on a falling tropical-forest tree. He explained that his wife was the artist and that they had commissioned someone to make up the shirts.

As late as 1999, then, some adamant critics remained unconvinced of any meaningful changes, especially on the long-term land tenure issues. And so far they are convinced that voluntary corporate responsibility is not the same as enforced accountability to policies of their own government.

These reflections about the plight of the small farmer were poignant and echoed the experience of small farmers everywhere. Once again, we noted the astonishing mixture of past and present, anecdote and suspicion, that permeates the rhetoric of activists intent on keeping the old history alive. It does not matter how things have changed; these people have real and deep grievances over company and government complicity in an economic model holding out little promise to many poor people.

Outside the Farm Gate

We knew that one consequence of the banana expansion was that land had become more expensive to buy for conservation purposes. An environmental spokeswoman confirmed this when we talked at the regional office in Siquirres, near the Tortuguero Conservation Area. Making it clear that she was speaking as a citizen in the banana region and not as a government official, she noted that Fundación Neotropica, a group that bought land and then gave it to the government, often found itself bidding against the big companies, which raised the price considerably. Although she did not volunteer the name of any specific company, she spoke with considerable authority about the relationship between the big fruit companies and the communities bordering the farms, since she had lived in the region before the banana expansion.

She confirmed a familiar charge that water leaving the farms had in the past been contaminated by agrichemicals and by the pesticide-saturated blue plastic bags used by all the companies to protect the banana stalks. The watershed drainage system carried this contamination through the Tortuguero Conservation Area (an area originally set up by the world-famous American naturalist Archie Carr) and eventually to the Caribbean, where there had been instances of turtle strangulation from the bags. Remember, this was the issue that reached the desks of banana bosses from children's letters as a result of the 1994 *Ranger Rick* article.

If a banana farm was next to a protected area, there were always more problems with poaching (including turtle eggs) and forest destruction. She went on to say that she didn't believe workers were convinced of the importance of conservation, nor was there a sense of community in the banana towns. Banana workers worked to get money and that was that. She felt that the companies could make a major contribution by incorporating environmental education for their workers.

The social situation in a banana town also was unfavorable, in her view. Children ran around unsupervised, much liquor was consumed, and there were many episodes of domestic violence. When the men came home from the farms, she said, they were exhausted and had no energy to devote to their families. The children had become real materialists, she added, and that was why they wanted to leave school. There were ways to instill other values, and the banana companies could make themselves part of the solution. They could do this at the community level, she believed, where there was a real chance to change behavior. While others in the area did not fully share these views, she spoke with conviction and with real concern, especially about the loss of values. Such regret about the trend of materialism in children is not unique to Costa Rica.

We discussed the matter at length with Alfonso Mata, a senior scientist at the CCT, the Tropical Science Center, a respected Costa Rican institution. Mata had been studying the impacts of banana farming on workers and on the environment. He had recently seen the results of the Better Banana Program on farms in Costa Rica and compared them to uncertified farms in Panama. In discussing the controversial aerial fungicide spraying to control Black Sigatoka, he was much more concerned about the moral and symbolic need to diligently avoid spraying around schools and worker housing than he was about the toxic danger. Mata had accidentally been under a crop duster at the wrong time and received a direct hit of the fungicide spray. He reported that he was extremely concerned and, as a chemist by training, calculated the density of fallout of the fungicide and the resulting "immision" at the surface. (The Spanish word "immision" is employed for the real quantities that reach the population once the pollutants have been diluted, dissolved, or dispersed in the natural media—water, air, soil, or living species.) He was relieved to find that the composition was dilute enough that he would not—he told us—be afraid to take a shower under it. Nevertheless, he strongly believed that the companies should take whatever steps were necessary to keep the spray out of the communities. The emblematic strength of that practice would be powerful.

In fact, given that Black Sigatoka control is not optional, certification requires strict attention to this issue. Chiquita had developed a timing and zoning approach that, together with a gradual expansion of buffer zones between schools, housing, and the farms, was a real improvement, but not a permanent way out of the dilemma.

Help for communities around the banana plantations is stymied by a frustrating quandary. The banana companies have been, perhaps understandably, reluctant to reclaim the high level of social responsibility that became increasingly burdensome as the high-profit years of paternalistic enclave agriculture came to an end. All have come to recognize that the previous kind of worker dependency is not sustainable. The Chiquita initiative to provide houses for worker ownership and childcare facilities suggests that companies may be able to reduce their costs while granting their workers more independence. Although in 1999 the company reported slow progress in this area, by 2003 the program had been reenergized. The government, on the other hand, saddled with the "structural adjustment" strictures conditioning loans from the International Monetary Fund, claims prerogatives but often of necessity withholds community improvements.[2]

Court of Last Resort

If the government cannot or will not act, the six-year-old national office Defensoria de los Habitantes, Costa Rica's adaptation of the ombudsman institution, is in a position to be a vigorous court of last resort.[3] Any citizen who believes that a government ministry is not enforcing the law and who is being damaged as a result can bring the case to the Defensoria. Using experts, it screens the case. If the agency, which is an arm of the congress and theoretically independent of any ministry, agrees to take on the case, it brings the matter to the attention of the relevant ministry and, if necessary, gets a court order to force the agency to fulfill its intended function.

We spent an absorbing afternoon with an associate of the organization

Technicians mix fungicides at an airport for spray planes. The moon suits are necessary because the men are working with concentrated solutions. Once diluted, the fungicides are reasonably safe to handle.

who had responsibilities in the area of "quality of life." With a degree in international environmental law, she proved to have a broad understanding of the global implications of the issues she dealt with. Passing through the antechamber, we noticed perhaps twenty complainants patiently waiting their turn, holding numbered slips dictating the order of the interviews. We sat down in a pleasant, shady courtyard where children played nearby. She said most of her dealings were with the Ministry of Health and the Ministry of the Environment. She noted at the outset that her agency was not well liked by either the government or the companies, but that in a poll taken in 1998 it was as popular with the people as the Catholic Church and the media. Its position as the most popular public body in the country gives the Defensoria its mandate and authority. She told us that her office remained engaged in holding the national government responsible for negligence by not fully addressing the problems of banana workers who had suffered sterilization from heavy use of the nematocide DBCP.

Between 1968 and 1979, some 5 million liters of DBCP had been injected into the soil around the base of banana plants through manual applications by thousands of banana workers in the Atlantic and southern regions of Costa Rica, as was reported in the 1998 *Cincinnati Enquirer* and elsewhere at the time. Most workers did not use gloves or protective clothing to safeguard them from absorption of the pesticide through the skin or by inhalation. As mentioned earlier, class-action law suits were brought against U.S. chemical and banana companies, but most were settled out of court though Costa Rican workers reportedly received a mere hundred dollars each.

The case would not die, however. On November 17, 1998, affected workers marched through the main streets of San José, demanding resolution of their claims. The marchers complained that the Costa Rican government had not allocated sufficient resources to attend to the affected population medically or scientifically. Not a single medical program had been specifically designed for this population, in spite of the fact that the state's responsibility had been confirmed in the final report issued by the ombudsman office in October 1998. The report had made a series of recommendations to the executive branch. The most important was the need to open negotiations with the companies so that the workers could receive a fair indemnity. Furthermore, the report had recommended specialized medical attention and pensions for the affected persons.

As a result, the government created an Inter-Agency Commission, composed of representatives of the National Insurance Institute and the ministries of health, agriculture, labor, and foreign relations. It carried out an eleven-month detailed investigation into the DBCP problem. In November 1999 (just a few months before our interview) the commission presented a summary report to the leaders of the protest that confirmed and expanded on the findings of the ombudsman. The commission's recommendations to the executive branch covered the three basic demands of the workers: pensions, specialized medical attention, and the creation of an indemnity fund. The agency associate explained that her organization had learned that it had to screen worker complaints thoroughly and objectively. For example, a worker complaining of sterility had to provide the number and dates of birth of all his children. Many of the complainants are illiterate, she explained, and they simply sign papers provided by lawyers, but the ombudsman office has to establish strong proof.

By 1999, six years after it was founded, the ombudsman office had received around two hundred complaints about the environment; noise generated from traffic outside hospitals; health fears about electromagnetic waves; and groundwater contamination from a variety of agricultural sources, especially coffee farms. Unions representing the banana workers had also come to her office with complaints. She mentioned one case involving workers for Dole and another involving workers on a Chiquita pineapple farm in the south, about water contamination from herbicides. She confirmed the sentiment of many that the government was not enforcing the laws with respect to the banana business. In her view the government does not want to "rock the boat" because bananas, coffee, and now tourism are so important to the economy.

The associate concluded that the certification initiatives have not stopped labor complaints. But on environmental issues she believed the companies were doing the best they could—not all the companies, but the responsible ones. She mentioned using biodegradable bags to protect the bananas and added that the companies know that the biggest buyers of Costa Rican bananas are European countries and that Europeans are concerned about environmental contamination. She said pointedly that if the buyers knew how the workers were treated, they wouldn't be so quick to buy. Yet, she repeated, at least some of the companies are very concerned about the environment. (At the time of the interview, Chiquita's social responsibility initiative and the 2001 labor agreement were still unwritten though being developed behind the scenes.)

When asked, the ombudsman associate confirmed that whereas many people feel the way she does about the social problems of the workers, others just think about the importance to the economy. They have a job, they have a roof. They don't care about the dignity of the workers. She had some reservations about Solidarity Associations, saying, "It's like a game we are playing. I have to see it in a few years."

We told her that Alfonso Mata, the CCT scientist, had volunteered that he had many problems with the government in Costa Rica, but that he didn't want his government to be like the one in the United States. He preferred, for example, the Danish model. Laughingly, the ombudsman associate responded, "It would be a dream."

The Workers

Interviews with workers outside the banana farms in Guápiles and Sarapiquí put a human face on worker issues. For most workers, labor and wage issues are more important than environmental aspects. When we spoke to a married couple, the husband working for Del Monte and the wife for Chiquita for five years, both confirmed that none of the banana companies did much for the communities outside the farms. A mechanic for one of the independent farms, who was a member of his local development association and school board, confirmed this; once he had sent a letter to all the companies asking for help in community projects, but had not heard from any of them.

We talked with two men manning the guard station of an independent farm that sold to Chiquita. They highlight a common situation for such workers. In September of the prior year, when other companies were doling out major pay cuts, the Costa Rican owner of the farm called a meeting and said his workers would get a 17 percent cut. Although he gave them the option of quitting and offered to pay them severance allowances, they wanted to keep working (as did most of the other workers). Since then, they said they had received a couple of cost-of-living increases and were almost up to where they had been a year before. They noted that there is always a flow of people in and out of the banana plantations, and that workers are always looking for a way to make a few more dollars. So they leave one place and go to another. The two men, both middle-aged, said that there had been positive recent changes on the farms. Before, no one wore gloves or protective clothing; now, everyone does. The owner was also planting along the riverbanks. Both thought the changes were worthwhile.

Because Costa Rica is a border country with a high level of development compared to the rest of Central America, it is one of the principal destinations for migrants from neighboring Nicaragua. With as much as 70 percent unemployment, Nicaragua has experienced extended years of war, exacerbated by a string of natural catastrophes that have hammered the country since its independence. In 1990 the peace process dampened the armed conflict in Central America. But over the years thousands of poor Nicaraguans have fled over the border. There are estimates that as many as half a million Nicaraguan nationals were in Costa Rica in 2002.

Many of them work on the banana farms. Some families are too poor to afford books or the necessary uniforms for their school-age children. In some cases, the communities help with food and clothing. We interviewed a Nicaraguan worker as he waited for a bus near Cariari. A handsome, muscular man with a dark complexion, he said he was a cutter and had been living in Costa Rica for seven years. He worked for an independent farm and wasn't sure which of the dollar companies bought fruit from his farm. He said wages hadn't been this low for a long, long time; he called them starvation wages. (When the economy was in better shape, we had been told that banana workers could make better money than a lot of white-collar workers in San José—bank tellers, or school teachers, for example.)

He said he knew of some people who were sick from pesticides, but he was vague and could have been talking about two or twenty workers. Yet, last year on his farm all the workers had started wearing protective gear. At first, he said, people complained a lot because they weren't used to doing so. The gear was hot and uncomfortable, but now they are used to it and he thought the protection was a good idea.

Later we asked Rafael Celis about the Nicaragua immigrant situation, among other issues. Celis, a resource economist, director of a private think tank, and author, linked the immigration to child labor problems on the banana farms. He theorized that although child labor is illegal in Costa Rica, the law had not been rigidly enforced on the farms because the government recognizes how important the banana business has been in buffering the effects of the migration. In Nicaragua, he explained, it is customary for children as young as five years old to work with their parents in the fields. Grandparents also pitch in. These cultural differences remain a thorny problem for advocates of uniform international labor standards and for the social accountability.

One could spend decades in Costa Rica and still not have a full understanding of its history and culture, but in just a few days we had seen and heard enough to suspect that the strong impact of communism—which had been a crucial influence on the labor movement throughout the isthmus for years—had weakened considerably. In fact, USSR-style organizing seemed to have proved disastrous to the union movement there. We wanted to confirm this possibility with an expert. At the National University in Heredia, northwest of San José, we spoke to Eduardo Mora, director of the environmental publications *Ambien-tico* and *Ambientales*. A serious and intense man, he sat down with us at a small table in the typically cluttered office of a professor. Behind him on the wall we noticed a large classic poster of Lenin. Behind us, in contrast, was a picture of a glamorous female fashion model.

Mora is a sociologist who has published extensively about the health of banana workers and the nature of their social organization. The old communist syndicalists still exist on some farms, he said, but they are weak. "Over the last sixty years, everyone detected an environmental crisis in my country—so did the leftists." In early postcommunist Eastern Europe, some said that former communists were like watermelons, green on the outside, red on the inside; but that would be too simplistic to describe the situation in Costa Rica. In his research Mora found that, contrary to popular American belief, the anticorporate worker stance was not driven by communist influence in Central America. Certain tenets of communism—as he put it, "moral communism"—still have some currency, but the communism of the former Soviet Union does not. In fact, as we were to discover, the principles behind workers' rights campaigns in Central America are now largely derived from the ILO conventions, with their history of joint industry/worker endorsement, and by the view that the theoretical trickle-down effect predicted by the capitalist free trade model had become a cruel trick played on the poor by the rich. In other words, you can be anticapitalist without being communist.

Early Assessment

In 1996, when Better Banana certification began to look promising to some of Chiquita's top management people, and when a number of their farms had been certified in Costa Rica, doubt still existed that certification could work throughout Latin America. The company was prompted to hire Conservation International (a fairly new but highly respected conservation organization with impressive scientific credentials) to look into the matter on the ground.[4] CI, as noted earlier, approached the Tropical Science Center to look critically at certified farms in Costa Rica and compare them to uncertified farms in Panama, a different country with a different history and a notoriously militant banana union. The CI-CCT study made a number of recommendations to improve the process, but generally found that in Panama, "if the Chiriquí Land Company utilizes Chiquita's environmental standards and implements the recommendations of this study, the company will achieve environmentally sound banana production."

Alfonso Mata, the CCT senior scientist (who, as we have related, felt so strongly about the moral and symbolic need to control the times and places of the aerial spraying even when highly diluted), told us that when Conservation International first approached his organization, the board was split on whether to engage in the study. Positive findings could condemn his institution in the eyes of anti–dollar-banana company campaigners and place the credibility of the center in jeopardy. Although badly divided, the board finally voted to go ahead, Mata said, and because of his own skepticism, named him head of the field investigations.

On the farms, Mata and his colleagues found concrete improvements. Having lived and done scientific research in Costa Rica and elsewhere for years, Mata was familiar with the traditional banana farming problems and came to the issues with a cool eye. He found that the improvements in practice were substantial and could be even better; with consistent improvements in some areas, they could be revolutionary. Though a leftist by intellect and temperament, Mata was quick to criticize the earlier union leadership in his country. The unions had promulgated the "myth of the *pobracito*," he said with some feeling. He believed that for years politically ambitious people had dominated the leadership. That was why the country was so open to Solidarity Associations (see Chapter 9.) Now freed from the influence of communism, the unions were becoming stronger, but it would not be easy for them to make inroads when banana workers were getting decent wages. From his obvious disappointment with his own government, one could easily infer that Mata too hoped for a government that could pass and enforce social policies promising more long-term protection to the workers than voluntary company policies, with their tenuous permanence over generations of management.

According to another source, it was also possible that the unions were finding recruitment a challenge among the new immigrants, many of whom were evangelical Christians, whose focus on the "hereafter" perhaps took their minds off the need for social revolution.

Major Players

Successive recent governments in Costa Rica have undeniably known and cared about natural resource and environmental issues. For example, the National Biodiversity Institute (INBio) is devoted to surveying the biological richness of natural ecosystems, pointing out gaps in the national networks of protected areas, and allowing the incorporation of knowledge derived from research in those areas into specific policies. Since the Rio Summit in 1992, Costa Rica, along with five other Latin American nations, has incorporated a healthy environment as a constitutional right and, in coordination with IUCN, has adopted a national conservation strategy. It has attempted to augment these national plans with sectoral action plans, such as those for forests.[5] However, critics charge that there is a lack of coordination among the various government branches, each being charged with enforcing a different aspect of the law.

It would be erroneous, in any case, to expect an EPA approach— even in Costa Rica with its deep dependence on ecotourism. We will, however, attempt to generalize about the nature of the forces driving performance on the banana farms. There are at least the following important threads to the tapestry.

The arm of the Catholic Church in Limón, as represented by Padre Vargas, was certainly important to the early strength of the protest movement against the dollar-banana companies. By far the majority of the Costa Rican population is Roman Catholic, and the church plays a fundamental role in the lives of the poor, many of whom work and live in banana country. Key priests, we were told, helped to organize the workers and helped them press their cases in the courts. The middle and elite classes in the society, with the leisure and knowledge to form campaigning nongovernmental groups, clearly augmented the role of the church.

The central part played by activist academics at the National University and other institutions like the Centro Agronomico Tropical de Investigacion y Ensenanza (CATIE) and La Escuela de Agricultura de la Región Tropical Húmeda (EARTH) needs to be recognized not only in raising the alarm about the danger of toxic agrichemicals, but also in advancing ideas about alternative agricultural practices. Some graduates of these schools, in fact, work for the banana companies, and some for other NGOs. None of these influences could have made much progress without an active and vibrant national press. Rafael Celis, in particular, was quick to single out how crucial newspapers were in the early days and weeks of the protest movement.

The trade union movement gained some strength from Foro Emaús and cannot be discounted, particularly in that it forged links with European activist groups and some elements in the region with international workers' organizations. All these forces were empowered by the rapid rise on the international scene of antiglobalization, anticorporate, anti-WTO groups, increasingly knit together by the Internet. Foremost among these are Banana Link in the U.K. and U.S./LEAP, a heavily networked Chicago-based workers' rights group. We confirmed our suspicions that environmental certification of Chiquita farms, as successful as it has been, must be seen in the larger context of the conflict between certain elites, who benefit from the fruits of the export economy, and the poor, who have been left out.

Most striking was how global the whole experience felt to us, especially the extraordinary reach of Europe into Costa Rica, even with its relatively Americanized surface culture. From development assistance agencies to small-farmer advocacy NGOs to supermarket banana customers sensitized by increasingly militant consumers and unions, the influence was palpable. Labeling movements and part-advocate and part-business enterprises like the Fair Trade movement draw heavily on this European history of protest.

Finally, it has been hard to ignore the reality that some of the issues raised by the campaigners are more tractable than others. Certification can and does provide a template for major performance improvements on the farms. But the Rainforest Alliance, like Chiquita, has also had to come to terms with the more stubborn trade, poverty, and agriculture issues that were so visibly raised on the embattled streets of Seattle.

7

Blue Bananas

In the last years of the twentieth century, northern Europe was the epicenter of a slow shift toward political consumerism. A few supermarket chains in the Scandinavian countries, the United Kingdom, the Netherlands, and Germany were beginning to understand the impact of changes in preferences in their markets brought on by politically motivated buyers who were transforming their purchasing practices accordingly. If the companies were slow to understand, activist groups were quick to give them a nudge. In the fresh food business, an alert grower could gain an advantage by anticipating this development. Chiquita and its rivals saw the trend coming. It was the beginning of a new relationship between the principal export companies and the large supermarkets in Europe and slowly in the United States.

To check out the changing market situation in Europe, we talked to supermarket managers, environmental activists, and labor representatives soon after the Brussels banana conference. Our first stop was Copenhagen. There we met with Mogens Werge, the manager in charge of environmental matters for FDB, the Co-operative Retail and Whole-sale Society of Denmark, founded in 1896 and now a huge grocery wholesale cooperative. FDB is part of a Nordic buying syndicate that includes outlets not only in Denmark, Norway, and Finland, but also in Italy. Preoccupied with some high-level management changes inside FDB, Werge nevertheless applied himself to the interview with candor. "Look," he said, "we know the banana companies are not angels. If the market wants blue bananas, they will provide blue bananas. If the market wants them environmentally grown, then they will provide them. They don't care if they're blue or ecolabeled."

Werge was referring to the co-op's decision to continue to buy bananas from Chiquita, and only Chiquita among the American banana rivals, to sell to consumers in thousands of European grocery stores. The decision had been triggered in part by scare headlines in 1998. The biggest Danish newspaper had accused FDB of buying bananas from a company whose practices had harmed workers in Central America. The charge was that chemicals used by Chiquita to control plant disease had caused sickness among workers and at least one death. Yet Chiquita had not been the only company implicated, and once again the media charge was based at least in part on episodes as far back as the 1970s. By 1998 all the dollar-banana companies were at various stages of improved environmental performance, but Chiquita was the only one to have pioneered third-party certification attesting to responsible treatment of workers and the environment. Under the circumstances, Chiquita had to be the leading contender for an FDB purchasing contract.

Werge said that in northern Europe grocery executives are accustomed to thinking in terms of the "political consumer." He harked back to the effective mass boycott when motorists shunned Shell gasoline stations to protest the threatened disposal of a mothballed oilrig, the Brent Spar, in the North Sea. Activists were fearful that leaks from the sunken rig would destroy marine life. The action convinced Shell to give up its plan. In another protest against French nuclear testing in the South Pacific, customers of one of FDB's competitors refused to buy French wine in order to send a message to the French government. "As a political consumer you have two choices," Werge commented. "Either boycott or make a positive choice of another product."

A Gradual Awakening

Later that week at Chiquita's European headquarters in the major shipping center of Antwerp, we had a chance to get the company slant on these issues from George Jaksch, director of corporate responsibility and public affairs for Chiquita Fresh Europe. Jaksch was the man in charge of quality assurance in Europe for Chiquita. Until quite late in the nineties, the big banana companies maintained a siege mentality on sensitive issues. Usually the lawyers or the PR people fronted for the companies, as had been the case when the *Enquirer* reporters sought interviews with management for their blistering attack. Our meeting with a top manager far from headquarters represented a significant change of corporate heart.

Jaksch was a pleasant, self-assured man who had had a long career in the banana business, both as an entrepreneur in Colombia and as a Chiquita executive in operations as diverse as Panama, Ivory Coast, and Europe. For a time he headed the sales operation for retailers in Germany, Switzerland, and Austria. Jaksch was also Chiquita's spokesman in Europe on environmental and social issues. "Environment is extremely important," he said. "As a European I know that the European public has a high level of concern and awareness of environmental issues."

He pointed out that the definition of "quality" in the banana business had expanded dramatically over the years. "But now," Jaksch continued, "the concept of quality extends beyond the cosmetic appearance of the product. The trend is toward health and safety, in addition to the environmental and labor conditions on the farms. The awakening is only gradually happening and is still extremely variable from retailer to retailer." The European Commission also promulgated quality standards that became effective in 1995, further institutionalizing size and color in the banana trade.[1] Typical language taken from these standards provides some sense of the cosmetic precision:

Every single finger of a package must be undamaged. This applies to damage that has occurred during the harvesting and processing as well as during the loading and unloading. The single clusters must not bear any sign of disease, physiological defects or rottenness. . . . The stems must neither be torn nor cut, nor may they become cracked during the packaging, transport, loading or unloading. . . . Whether the curvature of a fruit is abnormal depends on the curvature of the neighboring fruits within the cluster. In order to ensure safe packaging, all fruits of one cluster or one hand must have the same curvature. . . . The length of a banana is measured along the length of the curvature from the tip up to the beginning of the stem in the crown, it must be at least 14 cm long. In the context of size tolerances, though, a length of 13 cm is also tolerated.[2]

Jaksch said there is a big gap in environmental requirements between FDB's tough standards and those of other customers who buy Chiquita's bananas in Europe. In many places environment is not even a topic of conversation. "When we explain what we're doing, they say, 'That's very interesting but now let's talk about price.' It usually goes right down to the bottom of the agenda. I'm talking about some very major retailers. Others, like Migros in Switzerland, are keenly aware and supportive of our program."

Chiquita, however, was finding that European retailers tend to respond defensively. If an NGO undertakes, say, a postcard campaign protesting conditions on the plantations, it will ask Chiquita to provide it with records or even respond to the postcards. "So then there's a flurry of activity. But that's different from a coherent policy," Jaksch insisted. He also stated that supermarket chains in the United Kingdom are especially sensitive to conditions on the farms. Asda (a subsidiary of Wal-Mart), Tesco, and others actually visit the plantations in Central America. These big buyers do not require a particular standard, but they are deeply tuned into the issues, at least in Europe.

Jaksch then made the point that expectations are higher for Chiquita. "Given our leadership and our premium positioning of our brand, our customers would expect us to have an environmental program. In essence, certification provides us with the opportunity to do business but is not a contractual requirement with supermarkets," he said. Once having embarked on this course, Chiquita very quickly found a powerful incentive to keep it going. Certainly FDB, which had been threatened by the Danish union of stevedores and warehousemen to strike in sympathy with the Latin American banana workers' union if the workers' concerns were not addressed, made it clear to Chiquita that their relationship depended on Chiquita's performance in this area. "In a way we're grateful that they insist," Jaksch commented, "because it's motivated us to a high level of performance." Here was a clear signal directly from Chiquita providing one reason why the company was rolling out the Rainforest Alliance certification program. Fully 20 percent of its European market had been threatened.

"We had a critical dialogue with Chiquita," Werge had told us earlier in Copenhagen. "Either you improve the environmental and social conditions on your farms, or we'll find a supplier who will." Chiquita had to

ECO-OK on a Chiquita farm

take this threat very seriously. Already staggering from steep losses in market share as a result of the European trade restrictions, the company knew how important the Danish supermarket co-op was.

At the time of our meeting with Werge, a boycott by the union was still possible. Back when the media debate started, the union chief had told Werge that some of his members were saying FDB should stop distributing Chiquita bananas because, even with Better Banana certification, the company was still mistreating the union's fellow workers in Costa Rica. The union involved was the Danish Workers Union (SID), financed in part by workers' dues but also by the Danish government.[3]

"The union's job is to push us, but our mission is to try to push Chiquita in the right direction," Werge remarked. He believed that the

Rainforest Alliance should have gotten credit—not blame—for being willing to risk criticism by working with Chiquita. Most environmentalists were still resisting the idea of cooperative projects between environmental groups and multinational companies. IBIS, a Danish activist NGO, had been one of the most vocal critics of certification in this respect. In response to this activist skepticism Werge said, "We have an expression in Danish: 'Are you willing to get shit on your fingers or just stand on the sidelines with clean hands and criticize?'"

According to Werge, SID was involved in a project to assist agricultural workers in Central America. But the Danish government financed similar work by IBIS. Werge believed that it was IBIS that had fomented the headlines in the Danish press. "They need to make noise about banana problems so that the government will pay them money to go and solve them," he said. As Chiquita's Steven Warshaw later expressed it, "Nongovernmental organization is a misnomer."

In order to hedge its bets and bolster its standing in the European environmental community, FDB had engaged another Danish conservation group, Nepenthes, which sent biologist Dorte Nissen to Costa Rica to check on whether the Better Banana Program was credible. She had already inspected BBP-certified farms on an earlier trip and found them generally acceptable, but then Nepenthes came under fire from IBIS for accepting payments from the grocery chain.

Even though the Rainforest Alliance had teamed up with local non-profits in Costa Rica, it had been suffering from the "not invented here" syndrome at the hands of many Europeans who suspect that an American NGO cannot be trusted to certify the behavior of an American company. It was becoming increasingly clear to us that this bias was contributing to the persistence of the EU's restrictive trade regime for bananas.

Over a beer at a hotel on the docks in Copenhagen, we heard the Nepenthes side of this story from Nissen. A straight-speaking and clearly dedicated biologist with a strong environmental bent, she told us that FDB had given the watchdog role to her organization because Chiquita had been severely criticized in Denmark. Despite its huge size, FDB is the greenest supermarket chain in Denmark. "FDB knows that there is a political consumer developing here. These consumers don't boycott very often, instead they make their views known with their purse strings," Nissen confirmed. But boycotts are not out of the question. "If there's a multinational company that's doing bad environmental things, people could boycott their products. I'm very sure of that," she affirmed.

According to Nissen, Chiquita stepped on its own toes, too. Early in the certification process the company's marketing people underestimated the Danish consumer. She said that Chiquita's marketing was highly unrealistic. Nobody in Denmark believed its approach. "People here are too intelligent to accept advertising which claims that going onto a banana plantation is like going into a rainforest. Even the most stupid child could see that would be a lie. Maybe Chiquita's problems started because of that." (Chiquita had bought television time in Denmark comparing being on a Central American banana farm to being in a glorious rainforest.) "It was this advertising as much as anything else that put it in the spotlight," Nissen explained.

Certified for What?

Both Nepenthes and the Rainforest Alliance were accused by IBIS and others of not being sensitive enough or skilled enough to judge companies on workers' rights, as opposed to strictly environmental practices. The union admitted that establishing buffer zones of natural vegetation to protect streams from polluted runoff, reforestation, management of waste streams, recycling of pallets, and protecting wildlife from chemical poisoning were all within the expertise of both groups. Yet neither the Alliance nor Nepenthes, they insisted, had the credentials to certify that Chiquita was honoring worker concerns, especially child labor, hiring and firing practices, and freedom to organize—the core issues being championed by the Central American banana trade unions and the International Labour Organization.

When the Alliance conceived of Better Banana certification, conservation was its primary goal and that was very much in keeping with mainline environmental thought in the United States. But farmers' advocates in Europe consistently refused to allow a distinction between workers' rights and environmental protection. They hammered both the Rainforest Alliance and Chiquita endlessly on this point. And Chiquita, as a major agricultural employer in Latin America with a reputation from its United Fruit days of rough handling of workers, was exposed and vulnerable. Even Werge saw the distinction as ambiguous. "Is pesticide poisoning of workers an environmental or a social issue? The Alliance should admit it doesn't know and bring in the experts," he commented. In fact, over the years BBP had developed strong worker health and safety benchmarks for certification; it was only later that the BBP principles were expanded to cover a broader workers' rights agenda, as noted earlier.

The skirmish between activists and FDB seemed to have reached a manageable point, thanks in no small measure to the risky decision to certify taken by Chiquita years earlier. Werge summed up: "After a chaotic period, we landed in what I think is rather good cooperation with the union, though certainly not ideal. We're happy to have a critical dialogue with them, but we don't want to fight it out in the media—not because it would hurt banana sales, but because it's never nice to read about yourself, especially when it's not true."

As it turned out, Werge's fears of a possible strike were well founded. In a high-rise office on the outskirts of Copenhagen, Sune Boegh, a friendly no-nonsense man and the key union leader in SID for the banana trade, provided the union side of the story. "It could still be necessary to make some kind of boycott," he warned, confirming what we had already been told.

First, Boegh described what happened at the Brussels banana conference in 1998. The meeting had been partially funded by Danida, the foreign aid arm of the Danish government. It was at this meeting that the decision was made to target the dollar-banana multinationals, with a special focus on Chiquita. "Our goal was to initiate a dialogue between trade unions and Chiquita leaders in the U.S.—not the Chiquita managers in Central America," Boegh said. Among those present were Gilberth Bermudez, the Costa Rican leader representing COLSIBA and one of the organizers of the conference. "Bermudez spoke at the conference about problems with the workers," Boegh continued. "There was a representative there from FDB and agents representing Chiquita. We tried to push them into a dialogue with each other. At the time, Chiquita was buying television time in Denmark claiming that being in a Central American banana farm was like being in a glorious rainforest." This was something we had heard before.

Nothing happened between FDB and the unions until several months after the conference. Then, in February 1999, Boegh and SID's president met with the FDB union shop steward and others from FDB. SID gave the supermarket people twenty questions to ask Chiquita, ranging from environment to health and safety to workers' rights.

"I don't know if it's our pressure on the process or some other pressure, but I think it has been a good development that Chiquita and the banana workers' unions have begun to talk to each other," Boegh remarked. It seems that those early conversations were one of the turning points in Chiquita's long and stormy relationship with the banana workers. Three years after the conference, Steven Warshaw was to sign the historic agreement with the international agricultural workers' union chairman, Ron Oswald, agreeing for the first time to a basic set of standards for dealing with workers on all of their farms.

With respect to the political consumer around the world, Boegh noted: "You can still find Danish consumers closing their eyes, but there is a growing number of people who care. This has been a development of the last twenty years." Then he added: "If you look at foreign investment in the Third World you see some differences. If it is capital from the U.S., Canada, or Europe, working conditions are better than if it is capital from Asia, because Asians look at human beings differently. But the consumers in our part of the world are trying to generate their own pressure. Both the consumer organizations and we have some power in this process."

By agreeing to work together in the certification process, Chiquita and the Rainforest Alliance inherited each other's critics. It was an outcome that neither organization had foreseen, but one that would ultimately convince Chiquita's senior management that reaching out to adversaries is a crucial strategy. It would also lead to a set of dialogues between the Rainforest Alliance and European activists. Ultimately, the Alliance would have to find a way to incorporate the European focus on workers' rights into the certification process, and to deal with the social responsibility reforms that were already percolating inside Chiquita. The company's business rivals were probably delighted that the European campaigners had singled out Chiquita. The battle, it seemed, was going to be joined on a number of fronts far beyond the tropics—and probably far beyond the wildest dreams of crusty old Sam Zemurray, the Banana Man.

8

Agricultural Antagonists

Banana exports, 2001
000' tonnes (preliminary FAO estimates)

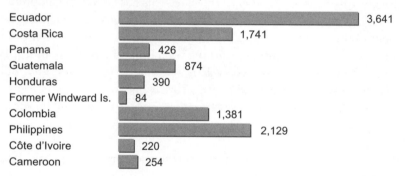

Ecuador	3,641
Costa Rica	1,741
Panama	426
Guatemala	874
Honduras	390
Former Windward Is.	84
Colombia	1,381
Philippines	2,129
Côte d'Ivoire	220
Cameroon	254

Banana imports, 2001
000' tonnes (preliminary FAO estimates)

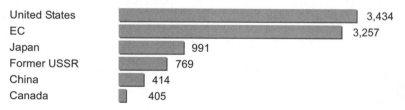

United States	3,434
EC	3,257
Japan	991
Former USSR	769
China	414
Canada	405

Source: Food and Agriculture Organization of the United Nations.(Shown in metric tons)

ven though the banana trade war between the United States and the European Union had been settled, other agricultural conflicts were fueling a growing antagonism around the globe. American brands were bashed relentlessly, and toward the end of 1999 the French flatly refused to buy hormone-laced beef from the United States. Washington struck back immediately with stiff tariffs on foie gras and Roquefort cheese. At just that moment, the gathering movement favoring small farmers gained a celebrated champion, José Bové.

An ex-Parisian who had learned English when his parents studied biochemistry at Berkeley, Bové led a small group of farmers in an August 1999 attack on a McDonald's construction site. It was situated in the small town of Millau in southern France, not far from the village of Roquefort, where the famous cheese is manufactured from sheep's milk. The franchise opened on time but all over France, McDonald's became a target. Later, Bové told Associated Press correspondent Mort Rosenblum: "McDonald's wanted to compromise by putting Roquefort on hamburgers. We told them that was like selling holy water in a sex shop."[1] Shortly afterward, having served three weeks in jail for vandalism, Bové joined his comrades on the streets of Seattle in the angry protests at the WTO meeting.[2]

Bové was one of several thousand people who were in Seattle in the fading weeks of 1999 to protest against the World Trade Organization, which aspired to accelerate free trade by launching a new round of trade talks. Agriculture in general is an ongoing trade conundrum as developing countries struggle to persuade rich nations to ease stiff tariffs and quotas set up to protect farmers in Europe and the United States.

Chiquita had been lobbying the U.S. trade representative since 1993 to persuade GATT, and later WTO, to rule that the European trade restrictions on bananas were illegal. In taking this position against protectionism, Chiquita surprisingly found itself allied with antiprotectionist forces throughout the Third World; as a large American multinational agricultural company, it symbolized in many minds a trading system that left out small farmers in favor of huge corporations and rich countries. Ironically, Chiquita, with headquarters in the richest country in the world, found its interests aligned with some of the Latin American governments whose developing economies were dependent on foreign exchange from the banana trade. These governments relied on the social stability that follows from the banana company's employment of thousands of workers.[3]

On the other side, some NGOs—and some governments, too—came to Seattle from Europe to persuade delegates to keep the complex and controversial banana trade system from unraveling. The EU banana regime was, in essence, an uncomfortable compromise caused by a major rift among European countries. Ever since the end of World War II, bananas have been a prickly problem in the genesis of a single European market. Before the EU finally came together, England, France, Italy, and Spain had established lenient import rules with their former colonies and existing territories in Africa, the Caribbean, and the Pacific (the ACP countries). At the same time, they levied quotas and heavy tariffs on bananas from Latin America produced on land controlled by the three big U.S. companies—Chiquita, Dole, and Del Monte.

Allied on the other side of the dispute were Belgium, Denmark, Germany, Ireland, Luxembourg, and the Netherlands. These countries had different stakes. Before the new regime was imposed, preferences for dollar bananas were not thwarted by quota, tariff, and license restrictions. Chief among these countries, Germany feared that quotas on Latin American bananas would cause the price of the fruit to soar.

The Banana War and the Small Farmer

If Europe was ever to consolidate its market, it was essential to reach a compromise on this thorny issue. The shape of the eventual 1993 compromise was foreshadowed by the Lomé Convention in 1975, involving forty-six ACP countries and Europe. Foreign assistance and trade matters were integral to the pact. Among the issues was the "banana protocol," in which Europe agreed that ACP banana exports to the European Community would not be any worse in terms of market access and advantage than they were at the time or ever had been. Subsequent revisions enshrined this principle, in effect allowing Britain, France, Italy, and Spain to continue providing trade protection to their former colonies. These ACP countries epitomize small-farmer economies—the same ones whose advocates stormed the streets in Seattle and set up websites on the Internet all over the world.

The acrimonious debates and fierce lobbying leading up to the final EU policies on bananas were tensely monitored by all the interested parties. With fingers crossed, the EU signed off on regulation 404/93 in February 1993. When the dust settled, the complicated new banana regime involved quotas, tariffs, and—the real bone in Chiquita's throat—a system of licenses allocated to importers of dollar fruit according to a formula designed to keep the ACP countries in the banana game, and to encourage new entries.

Under the complex and controversial compromise, imports of dollar bananas would be cut to a lower quota of 2 million tons a year, with a prohibitive tariff for any imports above that; fruit from traditional ACP suppliers (for example, Cameroon in West Africa and Saint Vincent in the Windward Islands) would be allowed tariff-free entry, with quotas based on exports before 1991. Stiff tariff rates would be levied against volumes higher than these amounts. A similar program was devised for imports from overseas EU territories with quotas roughly equal to those for ACP producers.

The vexing wrinkle for Chiquita, Dole, and Del Monte was a new licensing arrangement for importers of dollar fruit, the net effect of which was to force the American companies to buy licenses from traditional ACP and EU importers (such as Fyffes) and to create a market in the licenses in which some importers got rich for doing nothing. The bottom line for Chiquita was that it would experience years of enforced lower volumes entering its prize market and cost millions of dollars in lost potential sales. It was this kind of arcane protectionism that other developing countries, with the livelihoods of their small farmers at stake, would have argued against in Seattle if the rich countries had let

Realities in the Windwards

Bananas from small, independent farms are not necessarily produced in an environmentally friendly manner, despite the claims of some activists. Monocultures do not need to be extensive to require continuous application of agrichemicals.

In a thorough study of the banana industry on Saint Vincent, one of the banana-dependent Windward Islands in the southern Caribbean, Lawrence S. Grossman found that in one village, Restin Hill, 115 out of 144 families farmed plots of less than an acre, and only three families worked plots of more than 5 acres. Many factors have forced these farmers to apply pesticides: the pressure of buying associations; escalation, even locally, of the demand for spotless fruit; increasing international standardization of production methods; and pressure from multinational chemical companies. In *The Political Ecology of Bananas,* Grossman documents many instances of failure to wear manufacturer-recommended protective clothing; an incorrect assumption that only toxic chemicals smell bad; intentional and unintentional overdosages; and careless disposal. A representative of a Jamaican women farmers' group told us that in her country many small farmers choose to overapply chemicals in the belief that the crop production will be increased (unfortunately, a practice all over the world, the United States included). A Black Sigatoka–resistant banana strain could well be a major boon to the island nations of the Caribbean, and to Africa, India, and Asia as well.

In the absence of genetic breakthroughs and better practices in the field, Chiquita would probably argue that Better Banana certification moves the practices on Chiquita farms ahead of most banana cultivation by small farmers.

Thus, superior environmental management techniques may not be a reason to buy fruit grown in the Caribbean, but as Grossman points out, the U.S. victory in the banana trade war is probably very bad news for the small farmers on those islands: dependency over the years has left them with little to replace the banana trade. In desperation they could resort to planting marijuana or migrate in massive numbers to the United States and Europe.

them sit at the table. Actually, their interests were not so symmetrically aligned. When GATT tossed the hot potato to WTO, several banana-producing Latin American countries, including Guatemala and Ecuador, had already filed complaints under GATT about the unfair protectionist measures of the EU regime. The United States had supported this Guatemalan-led complaint in GATT and had invoked a section of the 1974 U.S. trade act on behalf of Chiquita, Dole, and Del Monte, charging that the EU banana regime distorted fair-trade policies under GATT principles.

Then, despite a number of unilateral declarations by the EU of its commitment to GATT and fair trade, the EU applied to GATT for exemption of the banana regime. To deflect some of the trade tension, it arranged a special deal. This "Framework Agreement" with four Latin American countries—Costa Rica, Venezuela, Colombia, and Nicaragua—allowed partial increases in the volume of bananas to be exported from these countries to the EU. Other Latin American banana-producing countries, such as Guatemala, Ecuador, Panama, and Mexico, claimed their Latin American counterparts had sold out to the Framework Agreement instead of maintaining a unified Latin American attempt to nullify the EU banana regime. Antidollar activists and business rivals in Europe could now portray Chiquita as an enemy of the small farmer and of human-scale agriculture.

This trade dispute and the plight of the small farmer were on our minds as we walked toward the center of downtown Seattle to get our press credentials. (At the time we were covering the Seattle event for our newsletter on corporate environment, health, and safety management.) We quickly discovered that the passive blockade of protesters linking arms in a human chain had effectively cut us off from the WTO meeting just as surely as it had mired everyone else, including delegates to the meeting itself. As we climbed up the steep Seattle streets seeking a way around the blockade, we were joined by a Frenchman representing Doctors Without Borders, the international NGO whose presence in Seattle underscored that group's determination to get urgently needed but patented vaccines and other medicines into desperately poor developing countries. Our press credentials proved useless that day, as we tried in vain to push through the melee. Tear gas drove us to seek refuge in a nearby restaurant, where we witnessed on television the escalation of the protests a block away. Along with the determined and colorful majority, the armored riot police, flaming dumpsters, helicopters, and bands of teenaged black-clad anarchists made for a scary beginning to an ultimately failed conference.

The Seattle moment of revolutionary theater was to be an early salvo in the growing antiglobalization movement and, as it turned out, only the most visible of the reasons for the failure to advance the trade agenda under WTO. The next day the police provided access to the conference site. In the corridors, representatives of a number of developing countries told us that they had been frozen out of meaningful discussions by the wealthy countries through a series of not-too-subtle administrative tactics.

Agricultural protesters, who cut their teeth in Seattle, were soon to grow into a serious worldwide movement. Almost a year later, more than five thousand farmers under the umbrella of La Via Campesina, a global farming body representing sixty-eight nations, gathered in the southern Indian city of Bangalore to protest corporate control of agriculture. The farmers, wearing green caps and scarves, shouted slogans against corporations selling genetically modified seeds, which they argued marginalize small farmers. "Our sovereignty over seeds, land, and water is slowly being taken over by corporations. When corporations take over our land, small farmers have no place to live or to go," said M. D. Nanjundaswamy, spokesman for the local farmers' forum who echoed the cause of the Third World Network activist organization at a public protest meeting. "Other sectors are incapable of absorbing the manpower that is involved in agriculture. We all will end up in hunger and death," he said. Rafael Alegría, Honduran coordinator of Via Campesina, publicly expressed "revolutionary solidarity" with Indian organizations fighting against the WTO regime and multinational corporations. "We are here to tell the world 'No' to the WTO, 'No' to the International Monetary Fund, and 'Yes' to our struggle. We are here to confirm our solidarity."

Feeding the World

After a storm of protest in the late 1980s and early 1990s, most activist groups had abandoned the cry for a general boycott on dollar bananas.

They reluctantly conceded how important the big companies were as employers and as providers of taxes to their host countries in Latin America. This uncomfortable reality did not deter them from criticizing the monocultures and the vertical organizational integration at the heart of the dollar-banana enterprises. Groups mostly in Europe urged shoppers to shun dollar bananas in favor of Fair Trade fruit which, they argued, leaves more money with the local farmers and is kinder to the environment and the workers.

Organic and other farm groups contend that the problem of monocultures (picture the vast seas of corn and grain in the Middle West) is that one crop, when it is extensively planted at the expense of other species, is vulnerable to disease. Consider a mixed hardwood forest invaded with elm disease. The elms suffer and die but the forest remains intact with its healthy oaks, beeches, and other trees. Monocultures, critics claim, set up a pattern of increasingly devastating human intervention. Pesticides in particular can endanger human health and degrade the environment. More intense applications can lead to chemical resistance, harnessing farmers to a pesticide treadmill. At worst, synthetic chemicals lessen dependence on time-honored techniques of soil improvement and change the chemistry of the soil. And specifically in the banana business, critics claim, the marked growth and concentration of large supermarket chains like Wal-Mart exacerbate the drive for those perfect bananas that can only be efficiently cultivated in chemically dependent monocultures.

Not everyone accepts these antimonoculture arguments at face value. One such skeptic is the outspoken Dennis T. Avery, director of global food issues at the conservative Hudson Institute in Indianapolis. Avery has written a book entitled *Saving the Planet with Pesticides and Plastic: The Environmental Triumph of High-Yield Farming* and he often challenges environmentalists. Here is an excerpt from a speech that he gave to commencement graduates of the College for Natural Resources at the University of California at Berkeley:

The fashion today is to worship the "natural" almost as did the Druids of ancient Europe. Such an irrational worship of the natural would be a desperate mistake for conservation. It would throw humanity back into direct conflict with Nature, a war Nature would lose. No matter how keenly the urban intellectuals of today revere wild things, humanity will never commit suicide to protect them—nor even permit its children to suffer malnutrition on their behalf. The well-educated 9 billion people we can expect on the planet in 2035 will understand good nutrition. They will have the affluence to pursue it, for their children and for their pets.[4]

Avery estimates that the world will demand nearly three times as much farm output in the year 2050 as it does today. Humans are already farming about 37 percent of the world's land area. An enemy of the wholesale adoption of organic farming—an approach with a solid niche in Europe and increasing in the United States—Avery calculates that with an organic farming mandate the world would immediately have to clear at least 10 million square miles of wildlands for green manure crops like clover and rye. Without continuing advances in knowledge and farm yields, "all the wildlands saved will be lost to the plow," he declares.

In 1998 Carl Pope, executive director of the Sierra Club, endorsed more research on high-yield agriculture (including bioengineered crops), because high farm yields will help save wildlife habitat and wild species. In a letter to the editor of *Philanthropy* magazine, which had published a controversial article by Avery, Pope wrote: "I strongly endorse the call for a renewed commitment to governmental and philanthropic funding of agricultural research, including research into conventionally bred or bioengineered new varieties of crops. A massive increase in such research is, as Avery argues, absolutely critical. Only then can the promise of high-tech breeding be combined with the social and environmental needs of the world."[5]

When asked about Avery, Dave McLaughlin said: "I have heard him speak on various occasions—and I think his message has value. It appears to be well researched, thought out, and takes a long-term view. However, I think he has a lot of detractors just by the way he hurls barbs at various other groups, organics and greenies. His essential message is about feeding people as world population peaks—and I have yet to see Greenpeace or anyone else rebut his concerns about food supply and population."

In response to Avery's views, some agricultural activists point to the failure of the Green Revolution of the 1970s, when new technologies in rice production were touted as a way to feed the world. Marty Strange, an agricultural writer, puts it this way: "There are good reasons why many of the funders (especially the Rockefeller Foundation) who Avery accuses of knuckling under to environmental extremists have not swallowed the high-yield farming bait. It is because the Green Revolution produced adverse impacts that deepened poverty and intensified hunger in some locales."[6]

The United Nations Food and Agricultural Organization, however, recognizes the importance of utilizing new technologies to increase food production. In a public meeting sponsored by the NGO Committee on Sustainable Development at the UN, the director of the FAO liaison office cautioned, "The challenge of the coming years is to produce enough food to meet the needs of an additional two billion people while preserving and enhancing the natural resource base upon which the well-being of present and future generations depend." Soil erosion is on the rise and is responsible for about 40 percent of land degradation worldwide. About 20 percent of irrigated land in the developing world has already been damaged to some extent by waterlogging or salinity. About 30 percent of livestock breeds are close to extinction. About 75 percent of the genetic diversity of agricultural crops has been lost since 1990.[7]

Additional fuel to this debate comes from the realm of genetic research. At the same UN meeting Robert Goodman, professor at the University of Wisconsin–Madison and a former executive at Calgene, Inc., raised some warnings about genetic modification (GM) technology. "We have less than ten years of significant field experience with these crops, and in fact only five or six years' experience with the widespread production of them . . . No responsible ecologist would be willing to make conclusive predictions about the ecological consequences, much less about the evolutionary consequences of GMOs [genetically modified organisms], with only five or six years of data." On the other hand, he estimated that billions of people have consumed GM foods worldwide and thus far there is no credible report of a health problem.

He noted that GM technology has been "almost exclusively in the industrial sector of agriculture." The technology, however, was within the reach of the normal plant breeder using basic tools such as crossing and mutagenesis to invoke variation. He believed that GM technology was not necessary for food uses of crops, though it was useful for making vaccines, pharmaceuticals, and plastics. GM technology does not have to be employed to achieve such goals as lowering the use of pesticides. Any plant breeder can do so with access to appropriate germ plasm— and funding.

Manipulating Bananas

There's more than a fifty-fifty chance that banana genes could be tweaked to make the fruit more resistant to disease, thereby drastically lowering the need for expensive chemicals applied by crop dusting airplanes—a favorite target of dollar-banana critics. An international consortium of scientists from eleven countries announced in 2001 that the banana would be the next major food crop to have its entire collection of genes decoded. Cultivated bananas have more or less been at an evolutionary standstill for thousands of years and lack the genetic diversity needed to fight off disease. Ancient farmers selected banana strains that were seedless and thus sterile, and grew the fruit through vegetative sprouting. Dave McLaughlin was familiar with the proposed effort. "We saw the decision to have the genome mapped. This group has been part of a World Bank effort initiated many years ago. I used to go to the meetings, but we saw little chance of success because of the massive bureaucracy and variety of interests on the table."

Its economic importance makes the banana an ideal candidate for sequencing, and so too does its vulnerability to disease. New strains are needed to resist the fungus Black Sigatoka, which also afflicts plantains, the starchy, potato-like bananas traditionally consumed in the developing world. The fungus, which attacks banana leaves and can reduce yields by up to 50 percent, is extremely costly to control with chemical spraying. Conquering Black Sigatoka would be an extraordinary victory.

Farmers in 120 countries grow an estimated 95 million tons of bananas annually, with 85 percent of the global crop produced on small farms for home consumption and local trade. Activists, especially groups concerned with the rights of indigenous peoples, have used these farmers to highlight the plight of the small farmer everywhere and the loss of ancient seeds.[8]

A spray plane

Genetic modification of the banana would seem to be a win-win situation. Dramatic reduction of chemical use and thus reduction of the danger and the cost would be a real breakthrough in the banana business. The small banana and plantain farmers—and the workers on the farms of the banana multinationals—would be the beneficiaries. But as GM adversaries will be quick to point out, and as Robert Goodman explained, nobody yet knows the unintended consequences of such genetic manipulation. European NGOs, in particular, are insisting on the so-called precautionary principle, which would make "Look before you leap" a public policy.[9]

Another warning was raised in the *New Scientist* that the two diseases attacking the banana monoculture, Panama disease and Black Sigatoka, could devastate all species of bananas so that in ten years' time there might be no more bananas. A group of French scientists at the Montpellier International Network for the Improvement of Banana and Plantain called the banana a sterile, seedless mutant that "hasn't had sex for thousands of years." The banana cannot breed, and its genetic makeup cannot fight off these fungi that cut fruit yields in half and reduce the

Farm practices emanating from BBP principles. Top left: *Sign advising no entry into an area that has been treated with agrichemicals. All certified farms respect the "safety intervals" set by the EPA and the BBP, during which workers must stay out of treated areas.* Top right: *A banana farm extending right to the edge of a river in Costa Rica, 1994. BBP has forced farm managers to back away from the rivers and plant buffer zones.* Bottom left: *Plastic bags waiting for the recycling truck.* Bottom right: *Solid waste trap filtering debris out of the service waters from a packing station. These traps, now at all packing stations, collectively keep tons of banana bits and plastic scraps out of waterways.*

productive life drastically unless there is continual application of fungicides. The dependency on the one variety, Cavendish, could result in a disaster similar to the Irish potato famine. For the small farmer who depends on the fruit as a daily food staple, this would be a human tragedy of major proportions. Whether genetic modification of the variety can build up resistance to these plant diseases remains to be seen.[10]

Nevertheless, Chiquita is working with a private research company to study the prospects of a banana plant resistant to Black Sigatoka.

Small field trials of GM plants are under way. The company gives several reasons why it believes its research is safe. First, since bananas do not reproduce sexually, they cannot threaten other plants through cross-pollination or animals through exposure to altered pollen. (This has proved to be a vexing problem with other plant material.) Second, the company is restricting research to include only the vegetative stage, not the fruit production stage of plant growth. According to its corporate responsibility report for 2000, Chiquita will base its final decision on whether to introduce genetically modified bananas on extensive field trials, but ultimately on whether or not "our customers and consumers have sufficient confidence in the health and safety of this technology." Clearly, banana resistance to Black Sigatoka through gene modification is not just around the corner.[11]

The increasing emphasis on food production for export rather than for local consumption is having a severe impact on agriculture generally, and the banana trade is inextricably linked to these tough issues. Small farmers are being displaced as land is increasingly turned over to production for export. Total national soybean area is estimated by Brazilian sources to have surged by 10–12 percent in 2002. The U.S. Department of Agriculture forecast for 2001–02 in the soybean area was a record 15.65 million hectares, up 1.68 million or 12 percent from the previous year. In Indonesia in 2001, a total of 3.3 million hectares were planted for palm oil, up from 2 million the year before.[12] Moreover, some argue strongly that small farmers are politically stabilizing and generally more desirable than armies of workers with no access to land. So far, these advocates have not been clear on how a transition from one to the other should occur; instead, they have primarily fought to retain existing small farms and fend off encroachment by industrial farming.

On the other hand, the word "vast" in terms of banana monocultures needs to be seen in perspective. In Costa Rica, for example, bananas occupy less than 1 percent of the country's land mass but employ more than forty thousand workers and constitute one of the top five industries generating income to the national treasury.

Since Chiquita, its competitors, its employees, and its host governments in Latin America seem to be locked into a historical model that depends on large-scale, vertically organized operations, critics will always put the company on the wrong side of the corporate–small farmer conflict. In terms of global food security, though, it may be on the right side—especially if one belongs to the Dennis Avery school of thought. For good or for ill, Chiquita and its rivals will always be factors in discussions of the small-farmer and global food security.

9

"Daylight Come . . . "

There is a good reason why there are so few big players in the banana game. Even with cutting-edge technologies and competent managers, uncertainties can play havoc with the most careful strategies. While weather and plant disease are crucial, perhaps no element in the business preoccupies managers more than labor discontent. With so many workers, Chiquita over the years has had much to contend with—and much to answer for.

If 1998 had been a wearisome year for Chiquita, then 2001 was the year the company emerged from turmoil into something resembling daylight. On June 14 of that year, in one of the most stunning social responsibility turnarounds in industrial history, Chiquita signed a landmark agreement with the International Union of Food and Restaurant Workers (IUF) and the Latin American banana workers' umbrella union group COLSIBA, a long-standing sworn enemy. With that act, Chiquita became the first multinational corporation operating in the agricultural sector to sign a "worker rights agreement." Chiquita pledged to respect worker rights as elaborated in ILO conventions, to work to ensure that its independent suppliers also respected those rights, and to address long-standing worker health and safety concerns.

The pact, entitled "Agreement on Freedom of Association, Minimum Labor Standards and Employment in Latin American Banana Operations," was signed by Ron Oswald, general secretary of the IUF; German Zepeda, coordinator of COLSIBA; and Steven Warshaw, then president and chief operating officer of Chiquita Brands International. Witnessing the signing was ILO director general Juan Somavía. Oswald described the agreement as "historic in the truest sense, meaning that it offers the possibility for workers and employers to seek a new basis for the resolution of problems in an industry which has throughout its history been highly confrontational." He added, "We hope that this agreement will serve as a model for relations generally between international union organizations and transnational companies."

U.S./LEAP, the American labor activist group that had been battling Chiquita and the other banana companies for years, issued a press release saying in part, "While Chiquita has far to go to implement the agreement, it has become the most transparent and engaged banana company with respect to its overseas operations, and one of the most transparent of any transnational."

"Transparent" because, in addition to signing the agreement, three months later Chiquita published its first corporate responsibility report, documenting in ninety-eight pages of unusual detail the company's financial, environmental, and social performance. A starkly candid section even gave workers' ratings of corporate performance—a number of them quite negative. One expert called the report "a masterpiece of clarity, relevance, and reliability."[1] The report set out the company's commitment to a social responsibility process that builds on the Better Banana certification program but goes far beyond it to incorporate a code of conduct for labor based on Social Accountability 8000 (SA 8000). This is a relatively new certification process developed by Social Accountability International (SAI), a New York-based corporate watchdog NGO. As Steven Warshaw told us, Better Banana gave Chiquita the confidence to move ahead with a social responsibility initiative. (See Chapter 10 for more details on SAI.)

Certain of the seeds of this remarkable agreement were sown, in 1998, at the banana conference in Brussels. COLSIBA and EUROBAN had convened this meeting to address what European NGOs were calling a crisis in the global banana business. It was there that the key parties came to understand that the old animosities had to be addressed for the sake of all the players.

Motivated in part by the Brussels conference, Chiquita met with a delegation of plantation workers' unions in Guatemala City in November 1998. All five of the Latin American countries with Chiquita plantations were represented. It was the first such meeting in close to a century. A second meeting was held in April 1999. At this point, the workers' unions presented a proposal for a Regional Workers' Rights and Environment Agreement as part of the international campaign to draw attention to Chiquita's labor record. Chiquita's ambiguous reaction was not as responsive to the notion of an agreement as many had hoped, but the company stayed in the game.

After months of intensive negotiations, COLSIBA, IUF, and Chiquita reached an international framework agreement. The accord was signed provisionally by the three parties on May 11, 2001, and formally signed in Geneva on June 14. The agreement applies to Chiquita's suppliers as well, and a joint union-company body will monitor its enforcement. In Costa Rica, which has the unique and controversial form of workers' cooperative known as Solidarity, a "Conciliation Procedure" was also negotiated, which recognizes that unusual type of association.

Just how important this labor pact would be to Chiquita (and therefore to its rivals) became apparent in the spring of 2002, when Human Rights Watch produced a report, "Tainted Harvest," based on careful research and on interviews of children, workers, government officials, and others. The researchers spent three weeks in Ecuador in 2001 collecting the data. In the report, the activist group accused all three American banana majors and Noboa (the big Ecuadorian family firm) of pushing a "race to the bottom" by not demanding fair treatment of workers on independent farms in Ecuador that sell to the companies. Human Rights Watch landed especially hard on the child labor issue. Researchers interviewed forty-five children and found that all but four had begun working on the banana farms or packing stations between the ages of 8 and 13. The report went further to accuse independent farmers of spraying children with pesticide from crop dusters. The authors asserted that despite national laws in Ecuador and their own codes of conduct, the companies have "disclaimed any obligation to demand respect for workers' rights" on the farms from which they buy bananas.

While noting Chiquita's rather minor role in Ecuador generally, the authors claimed that four of the children interviewed said they had worked on one of the farms from which Chiquita had purchased fruit in the past. The report acknowledged that Chiquita's vice president for corporate responsibility, Jeff Zalla, was the only executive from any of

Freedom of Association in Costa Rica—A Unique Arrangement

Only 15 percent of Chiquita's workers in Costa Rica belong to trade unions. In contrast, the majority of workers in the private sector have joined Permanent Committees, a unique form of labor organization with roots in the early 1950s, when local Catholic Church leaders conceived of an alternative that they hoped would promote social progress along with dampening a perceived communist influence inside some local unions. The concept gained real impetus in the banana sector in 1984 following a protracted strike in Golfito that caused a United Brands subsidiary to close down completely.

Once a majority of workers elect representation by a Permanent Committee, Costa Rican law requires an employer to bargain collectively with the committee, regarding wages, benefits, and conditions of work. The rank and file democratically elects the committee's leaders.

The existence of government-sanctioned Solidarity Associations side by side with Permanent Committees has caused some confusion. These bodies are not collective bargaining entities. Instead, they operate as voluntary private savings programs to which employers and workers contribute funds, which then become the property of the individual employees. Solidarity Foundations, as they are sometimes called, are not unlike credit unions. They provide funds for school uniforms and supplies, and enable members to get discounts on various goods and services.

As originally conceived, Solidarity Associations *were* entitled by law to bargain collectively. But the aspect of company contributions was more attractive than typical trade union benefits, and about a decade ago a barrage of international protests from the ILO and others caused the Costa Rican government to change the law, essentially separating Permanent Committees from Solidarity Associations.

Workers can join a union and also be members of Solidarity Associations. In the case of Chiquita's Costa Rican division, Cobal, 15 percent of the workers were unionized in 2000, 85 percent were members of Permanent Committees, and 48 percent were Solidarity Association members. Some critics continue to claim that freedom of association is being diminished, since an employer's contribution could be seen as an enticement away from union organizations. The company has agreed to review the matter.

the companies who had responded substantively to the charges. Human Rights Watch also claimed that although Chiquita buys all its fruit from Reybancorp, a company with each of its owned operations fully certified by Better Banana, Reybancorp in turn had bought bananas in the past from noncertified farms. Chiquita answered in a press release stating in part, "We have led the industry in adopting rigorous third-party performance standards in our owned farms, and have been working to influence our growers to adopt progressively more responsible practices."

Much more telling, since it came from a crucial third party, was the response of organized labor. IUF put out a thorough and uncompromising report of its own, praising Human Rights Watch for its tenacity and condemning labor practices in Ecuador. It urged Noboa and Dole, in particular, to institute immediate reforms. With respect to Chiquita, Ron Oswald, the IUF chief, noted:

> Chiquita certainly cannot escape its legitimate responsibility for what is happening in Ecuador. However, through its work with banana unions in the region and with the IUF, it has shown credible and practical intent to raise its standards and has been doing so across all of its operations. Del Monte apparently could not muster a reply to HRW and Dole resorted to "spin" which lacked both credibility and transparency and which sought to hide behind generalizations and dubious social responsibility awards and accolades. The IUF welcomes Chiquita's continued willingness to engage meaningfully with trade unions on precisely those issues that workers, through their unions, are exclusively able to improve in any sustainable or meaningful way. The IUF calls upon Dole, Del Monte and national Ecuadorian companies, particularly Noboa, to radically change their current policies and to meaningfully engage with trade unions wherever those companies grow bananas or buy fruit from local suppliers.[2]

This statement by Oswald had to have had a powerful effect on Chiquita's enemies and rivals. While keeping Chiquita's feet to the fire, IUF had at the same time honored its commitment to the company and sent a strong signal of support to Latin American banana workers. From Chiquita's vantage point, the social responsibility turnaround could not have come up with better early public results.

Rights and Demands

For the best part of a century, the banana story in Latin America has been a labor saga. Banana interests in the tropical zone gradually moved away from the paternalistic enclave model as governments became more democratic and unions became more demanding. Toward the middle of the twentieth century, United Fruit and its rivals began to see that sourcing fruit from independent farmers took a measure of uncertainty out of the business, leaving the independent owners to deal with workers and workers' rights. The companies had to balance this reduction in risk with a potential loss of control over supply.

It still made sense to hold onto land that might otherwise end up in the hands of competitors. As various governments began to talk about taking land back from the companies, the barons realized that the political muscle of independent farmers would work against any government move to nationalize the supply of fruit. In any event, the companies understood they were unavoidably enmeshed in the social dimension of the business.

As early as 1911, United Fruit in Central America employed some forty thousand English-speaking migrant Jamaican workers as plantation hands, including a number of managers, teachers, and clerks. The historian Richard P. Tucker notes that these were "uprooted men, bound only to industrial and cash structures, with no organic connection to the land."[3]

In 1917, the Latin American labor movement, encouraged in part by the Marxist revolt in Russia, was fighting both oligarchic governments and the entrenched imperialist American companies. But within two years communist parties had been established in most Latin American nations. The earliest workers' protests were against United Fruit's operations in Honduras, where a series of strikes was launched at the company's headquarters in Tela. This activity, in turn, triggered various responses from the U.S. government, bent on maintaining control over the Panama Canal. From this period on, the combination of economic dependence on fruit companies, increased gringo bashing, Marxist workers' protests, Central American government repression, and not-so-subtle U.S. government support of that repression, was the explosive reality. Conflict became violent. In Nicaragua in 1926, Cesar Augusto Sandino instigated a guerrilla-led rebellion against the government. He then focused on United Fruit, helping workers insist on better conditions and more pay.

A major strike in Cienaga, Colombia, in 1928 ended with a battalion of government troops firing on a crowd of demonstrating workers and killing a number of them (General Carlos Cortes Vargas, the battalion leader, admitted to thirteen dead). The massacre took on mythic proportions when Gabriel García Márquez invented the number three thousand dead in his book *One Hundred Years of Solitude.* Márquez later admitted that for dramatic purposes he had "to fill a whole railway with corpses" so he chose three thousand dead and the legend has now been adopted as history.

The episode came to symbolize United Fruit's severe treatment of banana workers everywhere—and became an exemplar of the laminated layers of truth and myth that have built up around United Fruit over its long history. Contrary to earlier theorists, recent research has found that the workers were not looking for a change in the existing system, but for a formalization of existing labor relations, many aspects of which were already codified in Colombian law.[4] It turned out that the real issue in Colombia was whether or not United Fruit could continue to remain at arm's length from the workers, and hence from union demands. As workers became more restive, imaginative tactics were brought into play to keep workers off the direct payrolls in order to avoid unions and their inevitable demands. In the 1928 Colombian "massacre," according to Marcelo Bucheli, just such a mechanism was at the center of the struggle. United Fruit had organized relationships with the workers using intermediary contractors, but the workers wanted to work directly for the company.

When markets in the United States and Europe virtually dried up during the Great Depression, lack of employment triggered worker protests against both the fruit companies and the governments, notably in Honduras. In Costa Rica, under the leadership of the Communist Party, a prolonged labor protest began with a massive strike against the fruit companies. Under the more liberal government installed at the time, United Fruit was forced to establish improved working conditions.

These episodes have contributed to the patina of myths that United

Fruit faced: constant labor unrest, strikes, and union busting. The picture is actually more complex. From the earliest days, when many workers were treated better than their counterparts in other industries, to the present, there were years of relative tranquility despite occasional strikes and violence. In some cases, goods available in the company stores were cheap and of high quality; in at least one instance, workers took unfair advantage of this largesse.

Communism and Labor in Central America

During World War II, banana operations were seriously hindered when German submarines patrolled the waters around Central America. Labor unrest continued to be a major factor once the war was over. The situation boiled out of control in Guatemala, where some forty thousand Guatemalans depended directly or indirectly on United Fruit and where 2 percent of the landowners owned almost three quarters of the land. In 1945 Juan José Arévalo, the new president of Guatemala (having risen to power after a dramatic leftist coup, the "October Revolution") instituted a series of reforms based in part on President Franklin D. Roosevelt's New Deal. Both land and labor reform were part of an ambitious overhaul of Guatemala's basic laws. Nevertheless, workers' actions against United Fruit were unremitting from mid-1948 to the spring of 1949. Major instability was to last through one more change in government, until the famous 1954 CIA-backed coup, aided and abetted by United Fruit, finally installed a government somewhat more congenial to the company—a story more fully told in Chapter 2.

Meanwhile, United Fruit workers in Honduras struck for higher wages. Standard Fruit (now Dole) workers followed suit. All told, some thirty-five thousand workers were involved and the banana business in Honduras ground to a halt. The American ambassador at the time blamed Guatemalan communists for the trouble. Then Hurricane Gilda hit Honduras in the last days of September 1954, devastating the banana plantations. At that point United Fruit fired ten thousand workers.

Banana workers in the 1960s were somewhat quieter. Relatively progressive administrations established elements of a welfare state and passed legislation protecting the rights of indigenous peoples. But starting in 1979, antigovernment Sandinista forces in neighboring Nicaragua toppled the Somoza dictatorship. Costa Rica became a fallback area for contra troops and anti-Sandinistas, largely under pressure from the United States, to whom Costa Rica was financially indebted. Adding more woe to the country, both the banana and coffee markets collapsed.

Throughout the 1980s civil war raged in Central America. It was during this period that the Costa Rican government, abetted by anticommunist elements in the United States, began to look for ways to discredit the union movement. Out of this situation the Costa Rican government, assisted by an arm of the Catholic Church in San José, developed the alternative to labor unions—the Permanent Committees. These could only have flourished in Costa Rica's unusual democratic and experimental culture.

A Dialogue Begins

On the ground in Costa Rica in 2000, we witnessed firsthand the beginning of a new international détente that would radically alter almost a century of conflict. Chiquita's regional managers in Central America had been cautiously reaching out to activists since 1992. But beginning in 1996, and intensifying in 1998 after the quadruple whammy (the Panama strike, the *Enquirer* series, the banana conference, and Hurricane Mitch), Bob Kistinger and Jeff Zalla in Cincinnati were beginning to forge tentative new contacts as well. This strategy was paying off, but often it is at this stage—after a company has started to make meaningful improvements—that activist attacks are at their most virulent. That is what seemed to be happening when we met with some of the union leaders in Costa Rica.

In March 2000 some movement toward reconciliation was already in motion. But tensions were still high and all parties knew that any wrong move could wreak havoc on the labor peace so close at hand. Our meeting with a SITRAP spokesman, his assistant, a banana worker, and, later, a woman who had recently been fired by Del Monte underlined some of the issues.

The individuals asked not to be identified. We were told that all of the banana companies in Costa Rica were antiunion, but that some were worse than others, especially Del Monte (Bandeco in Costa Rica). Despite the fact that in December 1988 Del Monte had signed an agreement respecting workers' rights to organize, the company fired "dozens of workers," all of whom were union members. Del Monte had apparently resorted to an age-old practice of terminating workers just before the end of a contractual probationary period, and then rehiring them at a lower wage.

At that point, the fired worker spoke up. She had been a packer for Del Monte for fourteen years and "four months." By contract, discharged workers are to be paid one month's severance for each year worked. But, she added, Del Monte had only paid her for eight months. Nobody who was fired got full severance, she said.

Although the Solidarity system is disliked by the unions and strongly criticized by some labor groups such as the ICFTU, the majority of Costa Ricans have accepted Solidarity as a progressive and viable alternative to unions. Many workers in other industries, including media companies, belong as well. Its presence will probably not go away until a popular labor leader emerges. The worker told us that she belonged to both a Solidarity Association and to SITRAP. In her view, the companies try to polarize the situation to convince workers that they need to belong to one or the other. She believed that Solidarity had set up antiunion committees to threaten workers into quitting the unions.[5]

What was the effect of Better Banana certification, we asked. Her reply was that environmental changes had taken place but they were minimal. Down deep, she thought the companies only really cared about bananas. Most of the changes were cosmetic, she said, adding that conditions in the packing stations were better, but she didn't like the loud music that prevented workers from talking to each other. Finally, she complained that when company managers pat a worker on the back for doing good work, it just pressures the others to work harder.

Generally, the complaints were against all banana companies, not just the majors, and their independent suppliers. Chiquita's name came up once, when the spokesman was trying to illustrate how the Costa Rican government was antiunion and procompany. He said that through an intermediary, the Banco Popular, the unions had finally arranged a meeting with then-president Figueres. They were forced to delay it, however, when a Chiquita representative called and demanded an immediate meeting with the president, bumping the union from its long-standing appointment.

A certain amount of money flows into the labor movement from development assistance agencies and NGOs in Europe. If it were not for help coming from European groups, just holding their ground would be very difficult for the unions.[6] All these workers believed that continued outside pressure from consumers would force the government to let the unions function.

Women Workers

The IUF-Chiquita agreement has opened the door for closer consideration of concerns that women workers have raised and with which both management and labor are grappling. Women work primarily in the packing stations, where they are usually responsible for the "desmane" (cutting the fruit into small bunches), the washing and culling, and the packing itself. Their complaints are fungus under the nails from having hands in water all day, varicose veins from standing long periods of time, skin allergies, and respiratory problems. They talk of ergonomic risks from repetitive jobs like putting stickers on the bananas and stacking the fruit in the boxes, and they complain of insect bites when the fruit first arrives at the packing station.

As elsewhere, most women carry a double workload because they are responsible as well for childcare, washing, ironing, cooking, and other household chores. Many are single parents. Psychological factors also come into play, chiefly sexual harassment from male workers and managers and lack of opportunities for advancement. In June 2000 an international women's banana conference called "The Smile of Nature" took place in Hanover, Germany. Women banana producers and workers from Latin America, the Caribbean, Ghana, and the Philippines came together for the first south-south effort by women themselves to address these issues. A representative of Chiquita also attended the conference, which was organized by the German Fair Trade organization Banafair.[7]

Management's Story

Let's turn now to the banana industry's side of the story. When we spoke to George Jaksch in Antwerp in 1999, he said that the banana industry had been under tremendous pressure. At the time, Jaksch was the senior director for quality assurance. In 2001 he was named director of corporate responsibility and public affairs for Chiquita Fresh in Europe. Cost issues are paramount. "If you've seen a union agreement, you know they can come to two hundred pages. They're extremely thorough. Because they describe individual work practices in great detail, they easily can become an impediment to changes in work practices. We absolutely need the flexibility to introduce new technologies. This often becomes a very difficult point of negotiation. The long-term viability of our farms is at stake." Jaksch said that in Costa Rica the company had been able to improve conditions and wages for the workers and at the same time introduce new technologies. It seems that reduction of production costs and improvements in the standard of living of workers *can* go hand in hand.

With respect to women, Jaksch said, "Women play an important role in our organization. Women staff many of our packing stations. Unlike many other tropical agricultural crops the banana business provides work for women and provides it year-round." He pointed out that many women are heads of families, therefore availability of employment for women is a social necessity. He added that the company had some programs solely for women. "For example, in Colombia, we have programs that help the widows—who lost their husbands due to the ongoing violence—to form their own small companies. Among other things they make the small pillows to separate bananas during harvest and have a small plastics recycling operation." Jaksch listed some other social initiatives, such as the AIDS programs that use peer counseling and programs and offer weekly doctor visits to the farms. "In the Ivory Coast and other countries we have instituted vaccination programs for children of our workers."

On one plantation in Costa Rica we saw the beginnings of a childcare program. We visited a woman who looked after several children in her home while her husband was working on the farm. Still, when we followed up on childcare facilities, the program appeared to be on hold. Chiquita's 2000 corporate responsibility report includes a discussion of the company's experience with issues raised by women. Only 10 percent of the workforce on company farms are women. Cultural norms in Latin America are part of the story, but jobs such as banana harvesting are physically too demanding for most women. Honduras is an exception. A high percentage of the workers in the packinghouses are women. The report notes candidly: "These jobs pay well compared to the field labor jobs of many male employees, but few woman advance to supervisory positions." The company has also discontinued pre-employment pregnancy and AIDS testing. Chiquita says that the testing was done to protect women from jobs that would be too demanding or hazardous and also to control the cost of health care, which is borne entirely by the division for all its employees. The change adheres more closely to corporate policies of nondiscrimination and employee privacy.[8]

Beyond women's issues are other labor/management conflicts. Dave McLaughlin gave us an example: The company had spent $500,000 on an automated process to move the bananas from the conveyor to the container truck. When we came to work one morning, we found that the whole apparatus had been trashed. "I think they did it because the motorized approach was an insult to their manhood. Young banana workers can be pretty macho guys," McLaughlin said. Ultimately, the company salvaged what it could and moved the equipment to another farm.

A Union Leader Speaks

In the spring of 2000 we interviewed Gilberth Bermudez Umaña, secretary general of SITRAP and founder of COLSIBA. A strong-looking, stocky man, Bermudez was generous with his time and unusually candid about his personal experiences, but he did not want to be quoted directly. He admitted that the labor movement in Costa Rica paid a very high price for its early affiliation with the Communist Party. Although he had been a member of the party himself when he was young, he said he had not been a leader nor had he played a role in the guerrilla wars in the isthmus. But he acknowledged that he had always believed the communists were fighting for social justice.

The years 1982–1987 had been crucial for the union movement in Costa Rica. Bermudez told us that the government had waged a determined campaign against the unions, especially since both parties were antiunion. The strongest repression came from Luis Alberto Monge Alvarez, who was Costa Rica's president from May 1982 to May 1986. Ironically, he was a former union leader. Bermudez said that during the

Monge administration the Costa Rican government was under tremendous pressure from the CIA to join the fight against the Nicaraguan rebels, with Oliver North flying clandestinely to a secret airbase in the north of the country. Bermudez asserted that Monge's government received more U.S. money from USAID and the Pentagon than the three or four governments before him combined.

Monge also aligned the administration with the military government in Honduras. In 1982, according to Bermudez, a press conference was held in the Gran Hotel Costa Rica. Present were the archbishop of Costa Rica, the minister of labor, American ambassador Francis J. McNeil, and Padre Claudio M. Solano from the diocese of San José. They proclaimed that they were going to recapture the province of Limón for democracy. The ambassador announced a gift from the United States of a million dollars to the archbishop's group to start a Christian crusade.

Bermudez said that the government's antiunion strategy included blacklisting, police storming of union offices, and a fierce propaganda campaign linking unionism with terrorism and with the communist movement throughout Central America. For Bermudez and his colleagues, it was a difficult time. He told us that Solidarity moved in, backed by enormous machinery mounted by the government and, he insisted, the banana companies. Padre Solano was a major player, according to Bermudez, and an early champion of Solidarity Associations. (He now directs the Escuela Social Juan XXIII, a training school in San José.) Bermudez said that Padre Solano linked the struggling union movement with Satan, atheism, and antichrist influences; he added that this caused him pain because his entire family is Catholic. Feeling the church was abandoning them, he and many other activists went underground. The companies would not hire union members, and those who would not join Solidarity were often hungry, Bermudez said. Then came a period of regrouping, during which he worked mainly at the grassroots level and eventually became a leader.

In the late 1980s, communism seemed all but finished. Bermudez commented that this was a period of strong emotion and discussion about what direction the labor movement should take. After 1990, the leaders designed an entirely new template for labor organization practice, and they had to do so with limited forces. Some union people had been fired from their jobs in the banana plantations and simply went on to other activities. Many union members did not join Solidarity Associations, but Bermudez admitted that some old activists did make the move. SITRAP is the only union that survived the storm.

By 2000, change was definitely in the air. Bermudez acknowledged that a majority of national producers want to reach some sort of solution to the problems. He said that a number of people, especially those in CORBANA (the government/industry banana trade group) agreed. At the very time we were meeting, he said that there was an ongoing dialogue between the Ministry of Labor, all the multinationals, CORBANA, and the labor unions. A high-level group was meeting every two weeks. Chiquita's head of the regional legal department attended; also attending were Dole and Del Monte and all of the big national producers. It was out of these meetings that sufficient common ground was found to lead ultimately to the signing of the pact between IUF, Chiquita, and COLSIBA.

Later, Dave McLaughlin described these meetings as the culmination of months of outreach work. Bermudez also claimed some credit. No doubt all the parties were beginning to see that the strife was not helping to convince the European Union or the activists that the banana republic days were gone forever.

At that point, labor peace still had some distance to go. The Costa Rican conservative group that included Padre Solano was boycotting the talks. Bermudez remarked that Solano probably was influenced by unrepentant cold warriors outside the Catholic Church. He referred to a conservative columnist and Costa Rica Libre, an anticommunist political group. All the same, he added that he and Padre Solano often had coffee together and regretted that the dialogues had pushed the padre into a corner.

Bermudez concluded by stating that he now believed there had been important changes in the thinking of senior management at Chiquita, although the innovations had not yet filtered down to the farms. He was basically very hopeful that agreement could be reached. When we asked him what would be the ingredients of a true agreement, he said he wanted to be optimistic, believing that human beings could change and companies could change. People could effect these changes. Banana companies had to make changes, and unions had to show they were not the enemy but that they could work together.

Great Trauma

The elements of the détente that we had witnessed in Costa Rica and Europe in 1999 and 2000 came together in the midst of great trauma in the banana industry. Both labor and management had to deal with the consequences of the worldwide banana glut. This inevitable connection between falling demand and labor strife, together with activist persistence, gave a new urgency to the tentative reaching out on both sides.

We have seen that a triggering moment for Chiquita came in February 1998, when banana workers in Panama went on strike (the first of that year's four traumas for the company). The union was protesting a violation of the collective agreement signed by Chiquita subsidiary Chiriquí Land Company and the union representing the Chiriquí Western Division's 4,600 employees, SITRACHILCO. The union charged that the company had left about 240 employees at the Pacific port of Armuelles without work by rerouting its exports through the Atlantic port of Rambala.

Other areas of complaint included the packing process required by the company, which had cut into workers' earnings, and the company's failure to pay workers for the ten minutes of rest to which they were entitled. When the strike ended fifty-seven days later, Chiquita estimated it had lost $10 million in forgone fruit sales. The company estimated it would cost a further $18 million to rehabilitate the plantations damaged in the strike.

The banana crisis continued into 2000, giving the workers increased anxiety. In August, Dole announced it would reduce purchases from fifteen Costa Rican independent producers by 25 percent and would not renew contracts with six independent producers the following year. Then in October Dole announced it would fire four thousand workers on its owned plantations in Costa Rica because of "lack of competitiveness." It later offered to rehire them with lower salaries and fewer benefits, as Bandeco (Del Monte) did in 1999. A Costa Rican newspaper reported that according to Dole, only 2 percent of the fired workers refused the offer.

When Chiquita announced in early October 2000 that it too would cease purchases of bananas from fourteen Costa Rican independent producers for the rest of the year, 7,200 banana workers had to deal with the potential unemployment. In November Chiquita announced that it would not renew contracts with six independent producers in 2001; these producers employed some two thousand workers. Earlier, Del Monte had said it would not renew any contracts with independent producers that expired at the end of 2000.

The multiple layoffs exposed the age-old dependency between workers and the companies, and lent considerable urgency to making peace between the warring parties. But behind the scenes, the big changes were already under way. Chiquita was preparing its corporate responsibility report, and the following year Chiquita, COLSIBA, and IUF signed the surprising labor pact. One can only imagine the risk and pressure involved.

New Dynamics

We have learned that the overhaul of corporate responsibility inside Chiquita started in 1998 with early contacts between Steven Warshaw and Business for Social Responsibility, but the transformation also involved the cast of players on both sides of the Atlantic. It had to have been a potent combination of business urgency, out-of-the-box thinking, and willingness on all sides to bend that carried the momentum along. When an unusual breakthrough like this occurs, big risks are taken on both sides. And whatever fragile trust has been created has to survive through the next generation of leaders and the one after that. All the parties need continued gumption and luck.

One crucial unanswered question is how the wild cards of social and environmental certification will play out as new factors on the battlegrounds of very old contenders—labor and business. Tensions are inevitable, since certification goals are not always in sync with the aspirations of either local managers or local unions. As social and environmental certifiers are fanning out into the banana plantations, a whole new dynamic has come into play. Dave McLaughlin put it this way: "Transparency is an issue we are still wrestling with. We do know from experience that a lot of times this transparency has been used against us, especially by the European NGOs. Where we recognize an issue and agree that improvement is necessary, they can turn it into a public condemnation rather than an exercise in honest dialogue. It's really like the Miranda rights: 'What you say can and will be used against you.' "

We find it necessary to report the obvious: that handshakes and solemn ceremonies are only a formal prelude to a grinding process on the ground. On May 27, 2002, just two months after Chiquita emerged from bankruptcy, some two thousand Honduran banana workers for the Tela Railroad Company went on strike at fourteen plantations and their packing facilities in Cortés and Yoro provinces in the north. The immediate cause was the firing of twenty-nine workers who refused to apply a pesticide identified as "chlorpyrifos" (or Durban), which the company started using on May 20. The workers said the chemical could cause vomiting, fainting, and sexual impotence. In addition to the rehiring of the laid-off workers, the strikers' twelve demands included a 10 percent wage increase and improvements in housing, schools, and medical care.[9]

After mediation by the labor ministry, Tela agreed to rehire the twenty-nine fired employees; the ministry also named a seven-member commission to determine whether the pesticide was harming workers. Yet a general assembly of union members on May 29–30 voted to continue the strike, adding demands for stable pay rates for workers in the packing facilities and insisting that the company show "respect" for a radio program carried by Radio Progreso called "Unionist on the Air." Honduran officials calculated that each day of the strike cost the government more than $600,000 in lost revenues. This was after Chiquita had laid off almost eight thousand workers when Hurricane Mitch destroyed much of its land in 1998 and the company had invested $75 million to rehabilitate 2,700 hectares damaged by the storm.

On June 4 the union leadership of the Tela Railroad Company voted to suspend the strike. According to one report, workers walked out of the session yelling, "They're paid off." But union leaders said they had to accept a labor ministry ruling that the strike was illegal. Citing the country's weak economy, labor minister German Leitzelar called on both sides not to "Argentinize Honduras." Evidently some of the leaders came to believe they had made a mistake in striking during difficult times in the Honduran economy. Suspension of the strike followed a twelve-hour negotiating session between the union and management. Management yielded to a number of demands on working conditions, and both sides agreed to continue negotiations on wages and other issues.

The walkout was called in spite of the pact between Chiquita, IUF, and COLSIBA. Apparently a representative from COLSIBA was at the table during the negotiations and may have been a factor in the settlement. Chiquita critics have said that the changes at the top in the company were very impressive, but that they had not yet filtered to the farm management in the tropics. In fact, it is astonishing that such radical changes could be dictated from on high in the first place, and they would certainly not have a chance of being implemented on the ground in the space of a few months.

When Chiquita published its second corporate responsibility report in November 2002, the company demonstrated its commitment to workers and produced a model for corporate reporting that is light years ahead of its competitors and, in fact, of multinational companies generally. In addition to a candid tracking of BBP certification plusses and minuses, the report presents in astonishing detail new steps and commitments on the social side. Highlighting especially the company's adoption of SA 8000 as the labor standard in its code of conduct, the report graphically sets forth the results of internal assessments against those social standards, using trained company personnel (see Chapter 11 and Appendix C). The company also invited independent observers to participate: German Zepeda, president of COSIBAH (Coordinator of Banana and Agroindustrial Unions of Honduras) and former coordinator of COLSIBA; and members of the Guatemalan labor rights monitoring organization COVERCO (Commission for the Verification of Corporate Codes of Conduct).

The assessments uncovered a number of significant gaps between practice and the SA 8000 standards. Some were acted on immediately (for example, the problems of Guatemalan farms were so intransigent that the company replaced the entire management team, which then arranged for union members to participate as observers of future internal audits), and a timetable was drawn up for addressing problems such as child labor, sexual harassment, and barriers to freedom of association.

With the company-wide commitment to SA 8000, the ongoing program with the Rainforest Alliance, the adoption of a set of core values, and the agreement with COLSIBA and IUF, it would literally be a miracle if the company ever slid back. Still, the transformation will never be total. Certain changes will prove daunting. The company and its certifiers will have to contend with integration of certain standards, and figure out a way to conform to international management system standards and various other gate-keeping phenomena being developed by European supermarkets. But in a very public way, Chiquita is clearly in the process of shedding its century-old reputation.

10

The Many Faces of Corporate Responsibility

To understand the roots of the social accountability standard, SA 8000, we need to retrace the evolution of business-to-business management system standards in the realms of quality assurance and the environment. We will then be better able to appreciate the rapid institutional changes that Chiquita and the Alliance experienced as their unusual relationship matured.

As Chris Wille and Dave McLaughlin circled each other in Costa Rica, campaigners throughout the world were beginning to demand that corporations be responsible not only for product quality but also for environmental and social conditions inside and outside the factory gate. They insisted that only a third-party assessment could validate such responsibility. What certainly influenced this escalation of expectations was the planning process for the UN Conference on Environment and Development (the Rio Earth Summit) in 1992. It was the first time that corporations were formally invited to participate. Moreover, the United Nations, as a body of governments, had been gradually letting nongovernmental organizations present their views in intergovernmental meetings. Years earlier at the UN Conference on the Human Environment in Stockholm in 1972, NGO "hippies" such as Wavy Gravy had been relegated to a former pig farm some distance from the official venue and corporations were barely visible. Only the formally recognized NGOs that included business groups were allowed to observe the debates and make formal statements at the discretion of the chairpersons.

Photo: A BBP auditor interviews the wife of a worker about housing provided by Chiquita.

Times had changed by 1992. The theme of the Rio conference was captured in the shorthand phrase "sustainable development." Although regularly maligned in some circles, it was painstakingly vetted to be meaningful in all major languages. The concept had been introduced in the eighties to address protests from developing countries that saw First World calls for global environmental protection as a conspiracy to prevent their development. Correctly understood, the much-debated concept captures the idea that true progress will only be possible when development does not happen at the expense of the environment, or environmental protection at the expense of development. Its overall aspiration is economic progress that does not jeopardize the future.

Economic development is the business of governments. But it can also be a consequence (intended or unintended) of multinational company activities. Business leaders from the United States and Europe knew they had to bring something significant to the table at Rio, even if it was only a work in progress. For some, the motive was certainly a sense of responsibility. For others, it was a simple case of "issue management": If the companies didn't take these social and environmental problems seriously, governments might well take matters into their own hands. Consumer boycotts were not out of the question, either. Fearing that the existing trade rules could sabotage environmental protection, and that the piecemeal UN apparatus for dealing with global problems was inadequate, some experts and governments also considered creating a World Environment Organization, or, alternatively, a combination of existing organizations with the strength to operate parallel to the GATT and its successor body, the World Trade Organization. Industry needed to meet this threat with some credible voluntary actions of its own.[1]

As early as the 1970s, business strategists focused on two forces that were to play a major role in the reshaping of corporate agendas twenty years later: first, the new but complex opportunities in global economics and trade, and second, the so-called quality revolution, which introduced the concept of certifiable standards especially between a large company and its suppliers. These twin elements came together to set a platform for industry's approach to the sustainable development imperatives that emerged from the Rio Summit. It led industry inexorably to voluntary third-party environmental and social assessments.

In this same period, governments and their experts on trade became convinced that the increasing complexity of the international economy demanded a more flexible, powerful international trade regime. During the final GATT round (the Uruguay Round, which began in 1986), governments completed negotiations leading to the creation of the World Trade Organization. In order to expand beyond slow-growing traditional markets and reach customers across national boundaries, especially in emerging economies, governments and corporations felt they needed a streamlined approach to the frustrating hodgepodge of trade-deadening protectionist policies around the world, and a legally binding dispute resolution mechanism.

Corporations Policing Themselves

On a course that was eventually to merge with the trade reform movement, the "quality revolution" was taking shape in the form of auditable standards. Driven first by problems in military procurement and later in new technologies like electronics, industry was seeking a practical way to add measurable consistency to production processes. An early example was the U.K.'s Central Electricity Generating Board and Ontario Hydro in Canada, which developed quality assurance standards for suppliers in 1969.[2] The crucial role of large purchasers is the heavy leverage they can impose on their suppliers. In 1987 the U.K. quality standard became an international quality standard (ISO 9000). It caught on rapidly. Certainly in the United States the trendy Malcolm Baldrige quality principles gave the standard a push.[3]

ISO 9000 was the first in ISO history to evaluate a management system instead of a product. It proved to be a bellwether for the first international voluntary industry environmental management system series of standards (ISO 14000).[4] Industry was able to bring, conceptually at least, to the Rio Summit its own contribution to sustainable development. The subsequent adoption was driven by a strong American reaction to an emerging European environmental management system standard with international ambitions, the Eco-Management and Audit Scheme, or EMAS.

IBM, with a position on the board of the American National Standards Institute, set out to convince the corporate board of directors at ISO that an international management system standard should be devised—but

Europe's Eco-Management and Audit Scheme

Although EMAS has elements in common with ISO 14001, its assessors are required to give stricter attention to the written policy during an audit. Moreover, the policy must be based on a review of all environmental regulations in every jurisdiction where the company operates.

EMAS in Europe was watched intently by U.S. companies, which were also feeling pressure to develop voluntary environmental programs. But both EMAS and its British counterpart BS 7750 contained elements that went against the grain for American companies entrenched in a much different, far more litigious culture than in Europe. In particular, EMAS demands that a company produce an annual report on its environmental performance. The report outlines the year's objectives and targets for its environmental management system and evaluates how well the company has performed against these benchmarks. Also, the firm is expected to note any future plans that could have a significant environmental impact.

The feature of the European standard most alarming to American companies was that it called for an external accredited verifier to review and validate the policy statement, environmental program, management system, and the audits themselves. This external verifier would be similar to the lead auditor from a registrar. Moreover, the verifier was required to confirm the company's periodic environmental statement. If a U.S. company were to fall short in the audit and that fact somehow were made public, lawsuits could soon follow.

EMAS survives to this date, supported largely by Germany. It now embodies ISO 14001, and the environmental management provisions are identical. And in 1997 ISO 14001 superseded the United Kingdom's BS 7750. Some American companies with operations in Europe have certified their plants to both ISO 14001 and EMAS.

one less fraught with liability concerns than the EMAS model. As a major international purchasing force, IBM had a big stake in the outcome. The ISO leadership no doubt also saw an opportunity to play a prominent role on the environmental stage.

Thus, in August 1991 ISO established a Strategic Advisory Group to assess the need for international environmental management standards and to recommend an overall strategic plan for creating such standards. An intensive consultation process resulted in which twenty countries, eleven international organizations, and more than one hundred environmental experts participated. What emerged was an elaborate array of rules that differed from EMAS in a number of ways. For example, EMAS can be more severe with a company's preparation of an environmental management system; any company with a past problem of land contamination must account for it. Under ISO 14001 the emphasis is on present processes. Under EMAS a preparatory review is part of the assessment. A company also has to analyze all inputs and outputs from processes at every site, including present, future, and past activities.

By early 1998 the United States had 83 sites certified to ISO 14001. The Ford Motor Company was an early mover, along with IBM. The worldwide total that year was 5,017. Inevitably, the standard had its critics. During the run-up to its final adoption, the core delegates in industry invited environmental NGOs to participate. The activists wrestled among themselves as to whether they should participate and ultimately decided they would not. The issue of NGO participation as the standards evolve is still a matter of intense debate.

Although representatives of the U.S. Environmental Protection Agency were involved in the meetings, the standards that emerged had a solid industry stamp on them. Journalists were never officially allowed into ISO drafting sessions. When an American journalist wrote about an ISO 9000 meeting and reported some unfavorable facts, industry reacted. Noting that several American journalists were covering the sessions, ISO issued a ruling that barred *all* journalists from every standards writing meeting. With few exceptions the policy still holds, but certain national delegations of the Technical Committee for Environmental Management accept journalists as members of their groups.

Some Third World companies complained that the new standard was, because of the cost, a nontariff barrier to trade that froze them out of the global economy. A number of small and medium-sized companies in rich countries took the same view. Most environmental groups felt that as a voluntary approach, ISO 14001 was better than nothing; it was a floor, but not much more than that. The main objection from these groups was (and continues to be) that the standard does not measure actual performance and, in the absence of a reporting mechanism, lacks transparency. In the banana business, for instance, an ISO-certified company could use virtually any chemical it wanted, a choice closed to Chiquita and others certified by the Rainforest Alliance.

However, even critics grudgingly accepted the reality that the changes a company had to make to its management procedures under ISO were a step in the right direction. Moreover, the clear pull on smaller companies from the bigger buyers created a powerful policing mechanism driven by self-interest. These forces could rival government regulations in rich countries and improve on them in many poor ones.

Certified companies could now confidently buy and sell products across borders in the knowledge that goods produced without environmental safeguards (and thus cheaper to produce) would not compete with goods produced in rich countries where stiffer environmental laws had already been enacted. Environmentalists in rich countries—the ones that could afford tough environmental regulations—began to recognize that their hard-won national campaigns would not automatically be undermined by a WTO trade regime that ignored these vexatious issues. Companies also felt that ISO 14001 helped to inoculate them from the rage of environmental activists when transgressions by a company in their supply chain, perhaps oceans away, came to light.

The People Part

With ISO's voluntary environmental standard finally taking root, companies started to feel that they could get back to business having put a major issue to bed—provided, of course, that they could master the arcana of ISO 14001 management imperatives. But small-farmer and labor activists were soon pressuring the big agricultural companies. Firms began to realize that the new environmental management standard might be necessary, without being sufficient, to earn a company the right to be called "responsible."

It was at the Rio Summit that voices were heard bearing the message that sustainable development had to put *people* back onto the global agenda. There was strong feeling that the environmental movement had sacrificed global social imperatives for the sake of natural parks and endangered species. ISO 14001 put a human imperative into the equation, with its implicit concern about balancing development needs with continual improvement of an organization's environmental management.

From the 1970s on, international conservation players such as the World Wildlife Fund and the World Conservation Union (IUCN) had been working with governments in the Third World. Together they had built theoretical walls and buffer zones around natural habitats, creating parks and reserves in an effort to arrest the quickening march of deforestation and lost species by preventing human encroachment. Sometimes locals were not even consulted in these grand conservation schemes, and the buffer zones had limited effect.

At Rio a backlash arose against this purist approach. Agenda 21, the main document that came out of Rio, was a political declaration of commitment and a program for action. Groups singled out for special attention were women, laborers, indigenous peoples, farmers, and youth—key players, after all, in the drive for economic and social equality. With virtual communities getting larger every day through the Internet, governments and corporations ignored these voices at their peril. Large and very visible multinational companies found themselves on the defensive once again. Some recognized that acceptance by the societies where they operated was, as many said, "a license to operate."

These issues heated up after Rio. In 1995, as the Rainforest Alliance was certifying Chiquita farms in Costa Rica and Chiquita was pursuing its head-on collision with the European Union on the trade regime, David C. Korten, a Harvard Business School professor, Vietnam veteran, and former advisor to the U.S. Agency for International Development in Asia, repented his capitalist ways and wrote a controversial book, *When Corporations Rule The World.* He argued that the classic business corporation no longer served society and as the principal engine of globalization was driving the world in a "race to the bottom." Korten made a persuasive case against the increasing concentrations of corporate power in fewer and fewer hands. Writing with the passion of a reformer and the vocabulary of an economist, he cited the fact that in 1995 the world's five hundred largest industrialized corporations, which employed only 0.05 of 1 percent of the world's population, controlled 25 percent of the world's economic output.

It is too early to know whether Korten's book will stand up to time, and there is certainly no sign that the mainstream corporate or Wall Street communities have paid it much heed. Still, it did help to energize the antiglobalization movement. A few progressive companies moved to address the so-called third bottom line of social responsibility. The Enron and subsequent corporate scandals in 2002 once again gave the responsible corporation a sharp new image of what it needed to be.

Voluntary corporate social reporting, or sustainable development reports, combining financial with environmental with social progress, were very slow out of the box. Many companies believed that supporting local scout troops or the local symphony was the most that could be expected of them. Moreover, a number of multinationals like Chiquita, operating in developing countries, had for years provided social infrastructure there—schools, soccer fields, health clinics, and the like.

There was much confusion even in well-meaning corporations about what was expected of them. As the nineties went on, calls for corporate codes of conduct grew more urgent. A number of companies responded, but as codes began to proliferate the need for a common template became clear. NGOs were skeptical (to say the least) of simple statements of high intentions. They demanded transparency through third-party audits, open reporting, and clear results on the ground. The treatment of workers by multinational companies was moving to the top of many NGO agendas. Sweatshops, used by name-brand clothing companies worldwide, were getting new scrutiny. Even though a cottage industry to develop so-called social metrics was building, agreement about what to measure in a company's social performance proved hard to reach, particularly in light of vast cultural differences across the globe. Child labor is anathema in Geneva, for example, but in some developing countries it is traditional and essential to help poor families survive.

Different Strokes

As we have seen, outside judgments on farm performance were not new in the banana business. Certainly by midcentury, jobbers and supermarkets were increasingly tough customers where ripeness, color, size, and curvature were concerned. Indeed, the technological ability of United Fruit and its rivals to deliver fruit closer and closer to the ideal would ultimately drive a war in which brand became a crucial differentiating factor. As the companies were able to deliver the fruit according to ever more exacting standards, the markets and middlemen stiffened standards too. It is truly amazing that this should happen with such a perishable fruit, especially one that required a long and hazardous journey. Few food commodities exported from the tropics to the table approach the banana in this respect.

Although Dole and Del Monte had decided not to seek Better Banana certification, in the midnineties (well after Chiquita had embarked on the Better Banana Program) both started the process of getting themselves certified to ISO 14001. Del Monte, by all accounts, lagged somewhat behind Dole. Chiquita for a time also flirted with ISO 14001. Banana boss Bob Kistinger told us in 1999 that he believed Better Banana to be a much tougher and more thorough process than 14001 because only the Rainforest Alliance program defines rigorous, uniform standards on the ground. In fact, developers of the industry standard intended that it have no requirements at all. It is a process standard that applies to the act of incorporating environmental aspects into management systems — the "plan-do-check-act" model of management. It says nothing about *what* must be done, only *how*. Dole's farms, by 2001, had undertaken some twenty-five ISO 14001 audits in five countries. Expenditures for training, improvements to infrastructure (warehouses, maintenance facilities, box plants, airports) and farms (packing plants, composting centers, reforestation), installation of treatment systems, and so on, needed to meet the environmental management system goals amounted to several million dollars.[5]

Today, Chiquita is moving along with a program in which, as contracts with independent growers are renewed, the growers are required to develop a program leading to Better Banana certification. The company is developing incentives for the independents to do so. The other majors generally do not require ISO certification from their growers because in most cases the cost would be burdensome.

These facts strongly suggest that competitive forces are driving environmental conditions generally upward in the banana export sector. It is also the case that the chief banana certifier for ISO 14001, SGS, is a for-profit business, not an environmental nonprofit organization like the Rainforest Alliance. Better Banana's on-the-ground indicators that are used to grant certification are public knowledge, and the Alliance's environmental commitment strongly supports the idea that certified farms show tangible results based on these indicators, even though the audit results are not made public.

SGS and other accredited ISO auditors are paid a fee for their audits that presumably is sufficient for the companies to make a profit. Certifiers under Better Banana are also paid fees for time and material, but the bulk of the support comes from private foundations and individuals. SGS and other ISO certifiers are accredited by officially recognized national bodies, whereas Better Banana auditors belong to a consortium of NGOs in Latin America—the Sustainable Agriculture Network (SAN). The Rainforest Alliance plans to be the accreditation body for SAN.

The principal difference between ISO 14001 and BBP certification comes down to this: With BBP, chemical use is monitored, water quality is tested, reforestation is measured, and numerous other changes are made to meet stringent and transparent indicators. Certification is made under the auspices of the Rainforest Alliance, which has to answer to outside funders, its own board of directors, the nongovernmental community, and any interested member of the general public. In comparing the BBP with other programs, Tensie Whelan, the Alliance's executive director, says: "We're the gold standard. Other conservation organizations work to stop conversion of land and prevent pollution, but without the accountability mechanism of certification. We apply these big concepts at the field level."

The Birth of Social Accountability

The twin ISO management system standards had laid the groundwork for industrial acceptance of a social accountability standard. But SA 8000, as it came to be called, did not initially come from the ISO business community. (At the time of this writing, ISO standard setters were contemplating development of an ISO *social* standard.) It came instead from the New York–based Council for Economic Priorities (CEP), a generally well regarded corporate watchdog organization which, under its founder and president, Alice Tepper Marlin, had for years been publishing report cards on big-company performance. Established in 1969, its mission was to "provide accurate and impartial analysis to evaluate corporate social performance and to promote excellence in corporate citizenship." This work brought Marlin into close contact with a number of big companies, and she gradually earned grudging respect from many of them.

In 1994 the International Labour Organization's International Programme on the Elimination of Child Labor invited a staff member of CEP to study what actions firms were taking to deal with the issue of child labor. The council convened a meeting with a range of corporations to discuss the matter. Encouraged by their response, between 1995 and 1997 CEP held further sessions in New York and London. It established an international advisory board of NGOs, trade unions, and the private sector to help devise a global standard for human rights. As many workers' rights representatives were included as company officials.

Out of these meetings, in 1997 (a year before the *Enquirer* series on Chiquita) CEP established an accreditation agency called Social Accountability International, the first of its kind in the world. With an advisory board of NGOs, companies, academics, and union leaders, its accreditation standard SA 8000 was modeled largely on the two ISO series of standards, but also drew on conventions of the International Labour Organization (see Appendix A).

A few points about SA 8000 are worth noting. The ILO has not formally endorsed the program, but is enthusiastic since it specifically references the seven core ILO conventions. SA 8000 departs in important respects from the two ISO management system standards In contrast to either of them, SA has on-the-ground performance stipulations on issues such as freedom of association, child labor, working hours, and discrimination, whereas certification to the ISO standards is based on third-party audits of management systems and whether the employees understand them. The SA 8000 standard allows NGOs to force reviews of certifications of companies when they suspect that practices run counter to ILO principles.[6]

Another difference is that SA 8000 was global from the beginning, in the sense that its principles were forged through an international group of representative stakeholders. In contrast, all ISO standards are developed at the start by national bodies whose preferences are arduously brought into line as a final international standard through numerous meetings of national standards organizations in many countries. Able to operate without the cumbersome ISO structure, SA 8000 got off the ground in about three years. Interviews with workers are more important than in ISO for obvious reasons.

Finally, unlike stand-alone elements in ISO, social accountability components are interlinked. For example, if forced labor is found to exist in a plant or farm, it is probable that other abuses such as working hours or disciplinary practices are also present. SA 8000 purposely does not cover the extractive industry, specifically exempting mining and oil drilling. Unlike Rainforest Alliance certifiers, and much more similar to the ISO approach, SA 8000 auditors tend to be private, for-profit certification businesses such as SGS or BVQI (Bureau Veritas Quality International), although NGOs can also be accredited. Companies like the approach, for it enables them to manage ethical relationships with other companies along sometimes long and complex supply chains. Two major companies that have been certified to SA 8000 are Toys "R" Us and Eileen Fisher.[7]

If the Rainforest Alliance has had to battle European activists over the not-invented-here hurdle, SA 8000, the other certification program with a strong American coloration, could not avoid the same early struggles. The story is told that in 1999, when Alice Tepper Marlin presented the SA 8000 approach to some European activists, they were unnervingly critical. But the program was definitely attractive to Chiquita's Kistinger. Since labor activists were blasting Better Banana certification for wearing social blinders, Kistinger was on the lookout for a companion approach that would do for Chiquita on the social side what the Rainforest Alliance had achieved on the environmental side. (By 2000, Dan Katz had been told by close observers that Chiquita was no longer vulnerable on environmental issues.)

While quite comprehensive, the early worker health and safety provisions in the Alliance's approach did not cover ILO precepts and therefore were not broad enough for labor activists. That meant they were not broad enough either for Chiquita executives, who had their eyes on the prolabor European consumer and the activists behind the trade regime. The experience with the Alliance had given the company the confidence it needed to reach out to yet another nonprofit watchdog group.

Kistinger says that at a meeting with Marlin in New York City in 1999, he found her ideas compelling, even opining that perhaps they did not go far enough.

The drive toward social responsibility had been gathering force inside the company even before the blast from the *Enquirer* in 1998. Steven Warshaw, the former president and chief executive officer, had resolved to salvage the company's reputation (among employees, at a minimum) through the bottom-to-top development of, in his words, "a disciplined path to corporate responsibility . . . not a public relations exercise, but a management discipline—a way to align ourselves around a core set of values, to build a sense of common purpose."

Warshaw assigned the groundwork of this effort to Jeffrey M. Zalla, then vice president for strategic analysis. Subsequently he became corporate responsibility officer and vice president corporate communications, and at this writing is vice president, treasurer, and corporate responsibility officer. We first spoke to Zalla at Chiquita's headquarters in 1999. A serious, deliberate man in his thirties, Zalla was cautious about going public with a program that was still in its early stages. A corporate steering committee had been formed, bringing in the senior managers who included Dave McLaughlin in San José and George Jaksch in Antwerp. The group was hoping to build one standard plan, plus a systematic implementation plan on the social side. For this concept to work, all the business units would need to sign on.

Zalla said that Chiquita was relying on Business for Social Responsibility to help the company look at a range of experience inside other companies.[8] Chiquita worked too with Instituto Ethos, a Latin American affiliate of BSR. Zalla and his colleagues on the committee were working toward the development of a code of conduct that defined a certain threshold of performance. He said they were also looking at how to engage outsiders and how to work effectively with other stakeholders – employees, customers, suppliers, NGOs, and other groups affected by the corporation in one way or another.

Earlier that day Kistinger had confirmed this new outreach commitment. He told us that Chiquita had made a firm corporate decision to go ahead and meet with the NGOs and try to engage them at every chance. "I tell my guys, go out and talk to the NGOs. See if you can get a commitment from them, that you can say anything you want, just call me up before you say it and let me give you my version. You don't have to use my version, but let me explain it to you."

In our interview with Zalla, he noted that at least some had not seen his company as credible, although he personally was proud of its recent past and of the direction it was taking. He agreed with Kistinger that the company should be much more forthright and that the Rainforest Alliance should be the model for the social side. "When you build in accountability for auditable third-party standards, you really drive the focus and performance of people throughout the company," he said.

Only a few months after the Panama union strike, and practically in the teeth of Hurricane Mitch, Chiquita hired SGS to do pilot audits. The idea was to help the company determine whether SA 8000 was a standard that was relevant and could work in its various operations. Over a period of more than a year, SGS certifiers looked at five of the company's owned farms in Latin America; at a banana puree joint venture; and at its canning operations. Auditors also looked at Chiquita's ocean shipping operations. It began to appear that SA 8000, with some adjustments and purposeful omissions, could work for the company.[9]

In the spring of 2000, Chiquita adopted SA 8000 as the labor portion of its overall social responsibility program, and over the summer a couple of dozen Chiquita assessors undertook internal audits of the company's seven banana divisions in Latin America. A contractor to SAI, together with Chiquita's own employees, trained these people. The assessors, using a forty-page questionnaire, reviewed records, interviewed managers, and inspected facilities. In consultation with management teams from each division, action plans were developed. All seven divisions then initiated their own action plans. These were the equivalent of a gap analysis, not an official SAI attestation. The idea was to develop an action plan so that they would be certifiable.

Chiquita Breaks the Mold

The next year, in the move truly raising the bar on corporate social accountability reporting, Chiquita published its first corporate responsibility report. Among the groundbreaking features was a graphically candid report of each division's compliance with SA 8000. It consisted of color-coded blocks showing which division was in compliance with SA

8000 in each category (child labor, forced labor, freedom of association, health, and safety). Green blocks indicate compliance; yellow blocks, an absence of required policy; red blocks, noncompliance in actual performance or not applicable. Although green blocks predominated, far more red and yellow blocks appeared than virtually any other company would voluntarily admit to (some 45 percent noncompliance; 43 percent an absence of required policy). Extensive footnotes explained most of the negative findings. For instance: "In Bocas, Panama, children, although not employed, were allowed to help clean up packing stations." Or: "The Honduras division lacked women in supervisory roles; required pre-employment pregnancy and AIDS testing, and did not hire workers over 40 years of age."

The practice of gap analysis by companies has become quite common. What is almost unheard of is the public nature of Chiquita's disclosure: such results are usually held in strictest confidence. Of course, the point is that the SA 8000 certification program was quite new and addressed some very difficult cultural issues. It is impossible to ignore the implied commitment behind this rare kind of corporate openness. Also interesting is that the practice in the divisions where the internal audits took place passed muster in a number of areas, despite the strictness of the criteria. One can speculate that some of the positive performance comes from practices hammered out in union agreements, and perhaps also because the Better Banana audits over the years looked at some of the same issues, especially pertaining to worker health and safety. Farm managers had already climbed part of the long hill ahead.

Among the companies being certified to SA 8000, Chiquita stands out not only for its size and brand, but also for its bold public commitment. Matt Shapiro, director of marketing at SAI, explained his experience in trying to sell companies SAI social certification "The analogy is the early days of financial accounting or quality auditing to ISO 9000," he said. In both cases, companies had at first been reluctant to undertake a complex new task, to try to move quickly to a standard, and to do it all publicly. "But they come to see that public accounting is now routine, and that getting to zero defects in a quality context is hard but doable." Then he admitted, "The secret to our success is simply the word 'trust.'"

11

Leveling the Label Field

When the Rainforest Alliance began the Better Banana Program, it quickly ran into rival organic and Fair Trade labeling movements in Europe. A few short years later, through a combination of success with Chiquita and tenacious reaching out across the Atlantic, BBP and the Sustainable Agriculture Network certification approach achieved—in the minds of many insiders—a rough parity with the more established labeling programs.

Social accountability labeling under SA 8000 certification has also gained increased respect. Leaders of all four attestation movements are currently working hard to find ways to accommodate to one another. The development is of considerable institutional importance since, in their various trajectories, these approaches emerged outside of any government intention or policy. All are mediating the relationship between the market and the industry. They are central to an understanding of the Chiquita story and of the worldwide trend toward customer-driven marketing.

The relatively rapid success of Better Banana and SA 8000 are major achievements when one considers the rich history and symbolism behind the two European labels. Fair Trade and organic movements reach back as far as the 1920s and draw on deep mystical, religious, and humanitarian symbolism to gain an undeniable emotive power. Perhaps

---

for this reason both labels manage to combine the strength of a social movement with a patently commercial face.

Until recently, Fair Trade and organic labels were largely content to live side by side in their comparatively different niches. As movements, both could campaign against the big banana companies on different grounds (small-farmer equity in the case of Fair Trade and "natural" farming in the case of organics), but neither had the broad environmental agenda nor the worker health and safety dimensions of the Rainforest Alliance approach. And the ILO focus was added later. Moreover, they did not have anything like the clout that comes from a relationship with a transparently transformed megabrand such as Chiquita. Nevertheless, all four programs had much to gain from the search for common ground.

Building on an early consumer base in Europe, the Fair Trade Labeling Organization (FLO) and the International Federation of Organic Agricultural Movements (IFOAM) had begun to make further inroads with some very loyal consumer groups there. The high recognition of their labels clearly was a headache for both the Alliance's Dan Katz and Chiquita's management, including Bob Kistinger. Katz mentioned to us that Fair Trade had taken an entirely different approach to certification than the Alliance. In marketing terms, Fair Trade worked for concept and name recognition with consumers first, then began to make connections on the ground with developing-country producers. The Alliance, on the other hand, did the hard pioneering work of getting things right on the ground first, then fighting for recognition in the markets. Chiquita people have asked, in essence, "What is Fair Trade doing to openly verify what it says it is doing?" It isn't a question.

There is no question that Fair Trade, as a social movement, if not a commercial strategy, continues to be a potent strand in the nongovernmental fabric, especially in Europe. Fair Trade, the political movement, essentially spawned Fair Trade, the commercial labeling enterprise. Proponents and operatives have not always made the distinction clear, and it is this cloudiness that has troubled critics. The movement has several antecedents, and as a result is a melange of business and small-farmer advocacy. The combination has been frustrating for Chiquita and its rivals, who felt this adversary was something of a wolf in sheep's clothing. It campaigned against the big companies as an NGO, but at the same time profited from the European trade regime.[1]

The Many Roots of Fair Trade

The first generation of Fair Trade ideas dates back to Oxfam, an organization that started as a group of Quakers and other church people in Oxford, England. It founded a committee for famine relief and assistance to victims of World War II. The group then turned its attention to poverty alleviation in developing countries. Oxfam people on foreign assignment originated the concept of selling the local handicrafts in England. The notion was to support indigenous people in Third World countries in a manner that embraced their dignity and skill. In 1965 Oxfam founded the "Bridge" program, through which handicraft artisans and farmers in Africa, Central and South America, and the Far East found a market in England for their products and a small but steady source of income.

Rapid growth in the "world shop" movement ensued: two years later, the Netherlands had 120 world shops and the model quickly gained popularity in Germany, Switzerland, Austria, France, Sweden, Great Britain, and Belgium. The first imported products were woodcarvings from Haiti. During the early years, only crafts products—the majority of them from missionary projects—were sold, mainly through missionary exhibitions, mail order, and Third World aid groups. (For more detail, see www.antenna.nl/fairtrade/fair662.)

Another crucial strand in the Fair Trade movement arose, again mainly in Europe, to combat the growing influence of multinational companies, especially those exporting commodities from the developing world. Vertically organized production and marketing companies like United Fruit were singled out, in the belief that their pricing formulas gave short shrift to small farmers. The influential Brandt Commission report in the eighties noted that multinationals (or transnationals) controlled up to a third of all world commodity production.[2] Although officially independent of governments, the Fair Trade movement often lobbies for policies and augments government efforts through its connections in developing countries. European development assistance agencies in the mid-1980s, for example, channeled considerable assistance funding into Central American countries in response to their explicit rejection of U.S. foreign policy there. The result was a palpable tension between American foreign policy and the European assistance agencies and their NGO allies, including Fair Trade and the international trade union movement.

Also in the eighties (remember, this was the time of the banana expansion), falling commodity prices contributed substantially to the increase in Third World poverty. The future development of these commodity-dependent countries relied on their ability to generate resources. To do this, they needed a substantial improvement in their export earnings. Successive rounds of GATT negotiations had largely failed to control the growth in protectionist policies. The Fair Trade movement sought ways to come to the aid of small farmers and family entrepreneurs in these countries.

Rich-country trade solutions were in disarray. The vacuum filled with a rising number of nongovernmental groups that followed the Oxfam example, based largely in northern Europe and known generically as alternative trade organizations (ATOs). The most successful second-generation ATO in Europe was the Dutch Max Havelaar Foundation. *Max Havelaar* is the title of a powerful 1860 novel that condemned "barbarism and corruption" on the part of Dutch colonial administrators in Java.[3]

In the fall of 1986 a national coffee campaign, conducted in the Netherlands by the Dutch NGO Solidaridad, went after the Dutch coffee roasters and their treatment of coffee farmers in Mexico. Capitalizing on growing consumer awareness, a group of organizations introduced the Max Havelaar labeling initiative to the Dutch market in November 1988.

Another Max Havelaar group was set up in Belgium two years later, and then one in Switzerland in March 1992. Again, the concept was to find ways that would help gain respect for the Third World farmer as an equal trading partner, and help these struggling people enter the general market by selling beyond the niche consumers who were loyal to the world shops. Large chain supermarkets became the targets, exactly the companies that Chiquita and its dollar banana rivals had set their sights on during the big expansion in the mid-1980s. (For more on this movement, see www.ecotopia.be/epe/sourcebook/3.6.)

Fair Trade leaders, after a while, began to pay attention to the workers as well. They promulgated a set of theoretical criteria to define performance on the part of small farmers producing under the Fair Trade rubric. The criteria were anchored in the Universal Declaration of Human Rights. Producers were to treat workers well, provide adequate housing, and pay wages above the national minimum. Moreover, workers would have the right to organize unions. Male and female workers were to be treated equally and paid the same wages for the same work. These principles were adopted by the ATOs when Fair Trade was a political movement and when the movement spawned its business phase; the principles were incorporated into promotional campaigns targeting consumers. In fact, in the Fair Trade system participating firms are not importers; they are more like brokers who seek out interested producers—often cooperatives. A procedure is established to certify that Fair Trade criteria are met. The brokers then approach traders willing to pay the higher prices needed in order to pay the producers the promised premium for following the criteria. The producer does not pay for the certifications; the costs are carried by the higher retail prices.

According to EFTA (European Fair Trade Association), the 1999 gross sales in Europe of Fair Trade importing organizations were estimated at approximately US$108 million, and the annual aggregate net retail value of all Fair Trade products in Europe exceeded US$234 million, of which US$189 million was food. Each country can authorize only one label organization.[4]

By the year 2000, bananas marketed as Fair Trade–certified sold in supermarkets in nine European countries. In Britain, where Banana Link plays a major role, the bananas are marketed under the label Fairtrade Mark. In most of Europe, the label goes under Max Havelaar or Transfair. In January 2000, the Co-operative Group Ltd. became the first supermarket to sell Fair Trade bananas in the U.K. According to Terry Hudghton, a corporate affairs executive with the consumer cooperative, Fair Trade bananas accounted for a significant proportion of total prepack banana sales. The concern has sold the equivalent of more than ten million individual Fair Trade bananas since the launch.[5] Switzerland is by far the largest market for Fair Trade bananas, which are distributed by large supermarket chains with a market share of about 15 percent. They cost about the same as conventional bananas.[6] Their success is attributable in part to the high consumer awareness of Fair Trade issues.

Fair Trade in Ecuador—A Different Story

Bananas marketed in Finland under the Fair Trade label are imported from Ecuador. A crew filming in Ecuador went to one banana farm expecting to find a dismal scene. To their surprise, the owner was the wealthy proprietor of a 340-hectare farm where 250 workers produce 40 percent of the Fair Trade bananas sold in Finland. The team found that the Fair Trade subsidies had resulted in an improved health center, increased school materials, and a newly employed teacher. But the owner had used the Fair Trade surcharge passed on to him from the Finnish consumers to enroll his workers in the Ecuadorian social security system—actually a statutory obligation of the employer under Ecuadorian law. The workers proved to be paid less than those on a nearby farm, although FLO marketing literature claimed that Fair Trade workers receive wages 20 percent above average. The television team also found discrimination in pay between men and women. The producer asked whether charity to a chosen few was any kind of answer to the social predicament of Ecuador's 300,000 banana workers.

For contrast, the TV crew visited a Favorita farm owned by Reybancorp, the second-largest producer in Ecuador. (The company supplies bananas to Chiquita for export to Finland after certification by BBP.) The crew reported that the Favorita farm resembled a tropical garden in which flower plantations alternate with bamboo copses and patches of rainforest. They then concluded that the only natural rainforests in Ecuador's coastal region were on this farm, in and among the banana plantations. The television program also charged that the Finnish Fair Trade Association was using a significant portion of the surcharge for its own expenses instead of passing on the money to the workers in Ecuador. This accusation forced the association to discontinue importing bananas from the wealthy Ecuadorian until an investigation could be made.

Dave McLaughlin said, "Personally, I believe the goals of the Fair Trade movement to be noble—helping the small farmer compete and trade internationally. However, as they have grown, their motivation may have shifted toward fulfilling commercial considerations (consistent volumes, quality, etc.), and so the lines between their original purpose and their commercial motivation may have become blurred. I do take issue with their criticisms of Chiquita and the others—they themselves are being found in violation of their own standard. Net, net on this issue—transparent verification is a big matter. A lot of small farmers could be hurt in this process, which would be terribly unfortunate."

Some of these problems were foreseen by FAO in its 1999 analysis of organic and Fair Trade bananas. It finds, in part, that Fair Trade needs to spend more money on communicating with the consumer and retailer. Unlike organic certification, which has a scientific basis, Fair Trade has not been able to forge an international consensus on its "softer" social criteria. The FAO study concludes, "Current efforts by the Fair Trade Labeling Organization to harmonize criteria and labels are important if the visibility and the credibility of fair-trade bananas are to be enhanced."[7] Some question whether Fair Trade should be working with bananas at all. They argue that propping up small farmers trading a volatile and perishable commodity is not really sustainable in the long run. Instead, perhaps it could help small farmers turn to some other crop. In this sense coffee, with its legion of small farmers, is a perfect Fair Trade vehicle.

It would seem that the Fair Trade and dollar bananas could coexist peacefully. A great deal of useful and important work could be achieved by this movement without the bashing strategy, and there are signs that this is already happening. But it is hard for a conscientious consumer to resist the word "fair" on a label.

The Mystical Roots of Organic Agriculture

The Austrian philosopher Rudolf Steiner is the mystical father of organic agriculture. He believed that to live in harmony with nature's equilibrium, humans must understand the need for a balance between the spiritual and material sides of life.[8]

Interestingly, two years after Steiner's lecture in 1924 on the benefits

---

Excerpt from a 1924 Rudolf Steiner Lecture in Dornach, Switzerland

There have already been agricultural conferences in which the farmers have said: Yes, the fruit gets worse and worse! And it is true. But naturally the farmers haven't known the reason. Every older person knows that when he was a young fellow, everything that came out of the fields was really better. It's no use thinking that one can make fertilizer simply by combining substances that are present in cow manure. One must see clearly that cow manure does not come out of a chemist's laboratory but out of a laboratory that is far more scientific—it comes from the far, far more scientific laboratory inside the cow. And for this reason cow manure is the stuff that not only makes the roots of plants strong, but that works up powerfully into the fruits and produces good, proper protein in the plants that makes man vigorous.

---

of using cow manure in farming, United Fruit established the Lancetilla Botanical Garden and Experimental Station in Honduras. Lancetilla was a key factor in a program to diversify United Fruit's plantings in Central America. Ultimately, it led to the promotion of new cash crops in Honduras, such as the African oil palm, eucalyptus, and teak.

The Steiner approach was elaborated in a practice of "natural" soil improvement in farming. So-called biodynamic agriculture was developed toward the end of the 1920s in Germany, Switzerland, England, Denmark, and the Netherlands. The theory implies a philosophical idealism that leads to specific cropping methods linking agricultural activities to the lunar and astral cycles. In England after World War II, this movement (later taking the name Soil Association) was dedicated to the restoration of humus and soil fertility to their original places in the biological balance. In organic agriculture, disease, pests, and soil fertility are mainly controlled through crop rotation designed to vary the crop demands on the soil from year to year, to break pest and disease cycles, and to alternate crops.

In the U.K. today, as Steiner taught years ago, in organic agriculture, soils are enriched by livestock manure, composts, and leguminous plants, principally clover. Pest and disease problems are reduced over time by increasing diversity of crops, using resistant crop varieties, and timing cultivation appropriately. A small number of natural treatments may be used, often requiring permission from a certifier. Weeds must be controlled without use of herbicides. One of the pesticides allowed in the U.K. is naturally occurring rotenone. The remainder (soft soap, sulfur, and copper) are synthetic but traditionally used in organic farming. Rotenone and copper are allowed under restricted circumstances only, on a case-by-case basis. "Organic" does not mean "no chemicals," it means plenty of hands-on work to minimize harm and to mimic natural processes as much as possible.[9]

The protest movements in the 1970s, with their alternative lifestyles and "back-to-the-earth" subculture, gave a major boost to organic farming in the United States. (Everything in those days seemed to be gnarly and hairy, including a growing number of organic vegetables.) Formal structures began to take shape in this period. In England, the Soil Association created a logo and promoted it as an attestation backed by performance criteria and quality controls and implying a legally binding guarantee for the consumer. In France, organic farmers' syndicates organized into federations such as the FNAB (Fédération Nationale d'Agriculteurs Biologiques). It was in this period, finally, that the principal national organic farming organizations in countries around the globe formed IFOAM, the International Federation of Organic Agriculture Movements.[10]

Small-scale banana farmers were virtually the only producers of organic bananas until quite recently. But large-scale plantations, notably in the Dominican Republic and Ecuador, have recently started exporting organic bananas. After some years of pilot testing, Dole entered the organic market in 2001 with imports into the United States from Ecuador and Honduras. In the same year, Fyffes joined the organic trade with imports into the United Kingdom. Even during its difficult reorganization, Chiquita too began trials.

Imports of fresh organic bananas into the United States and Canada were estimated at 16,000 tons in 1999 and at more than 23,000 tons in 2000. The annual growth rate was 33 percent for 1999–2000. Still,

Estimated Fresh Organic Banana Imports per Year by Region / Country

	Region/Country Imports (000 metric tons)					Annual Growth (%)	
	1996	1997	1998	1999	2000	1998–1999	1999–2000
U.S. and Canada	8	—	13	16	23.5	25	50
Western Europe	—	10	13	23.5	42.5	80	80
Japan	—	2.5	3	5.5	9	80	70
Total	—	12.5	29	45	75	55	65

SOURCE: FAO, Committee on Commodity Problems, Intergovernmental Group on Bananas and on Tropical Fruits, Report of the Second Session, and idem, The Market for "Organic" and "Fair-Trade" Bananas.

organic bananas account for only 0.5 percent of total banana consumption in the United States. FAO experts suggest that health concerns are a factor in this growth. Some of the larger cities have seen a noticeable but not exceptional move toward organic produce in general.

The Organic Market

Estimated imports of organic bananas during recent years in major markets are given in the accompanying table. The annual growth rates have been very high, but volumes of organic imports are still small.

In 2000 the Dominican Republic was by far the largest exporter of organic bananas at around 44,000 tons—an 80-percent increase over 1999. These shipments accounted for more than half of the global supply of organic bananas in that year. Mexico was next, exporting only about 20 percent of the Dominican Republic's volume. Both Colombia and Ecuador were becoming more important players.

FAO notes that food scares have been a major cause of the rise in sale of organic products. Organic bananas seem to taste better to some consumers, even though scientific evidence has yet to confirm that eating organically grown food is any healthier than eating produce grown in any other way.

Options of the Independent Banana Farmer

Generally speaking, an independent farmer can sell his fruit to one of the big five exporting companies (Chiquita, Del Monte, Dole, Fyffes, Noboa) or to one of their subsidiaries. He can decide to make arrangements with a local exporter. He can join a Fair Trade network, which seeks to pass directly to the farmer a premium paid at the supermarket, thereby, in theory, partially offsetting the imbalance imposed by the predominant free trade regime. Finally, he can join the growing number of farmers who choose to maximize "natural" inputs in the growing process and become certified as "organic."

If the farmer decides to sell to a Chiquita subsidiary, increasingly he must first become BBP certified by the Sustainable Agriculture Network. Chiquita at times may motivate producers to get certification by offering a small but significant price premium as well as valuable technical assistance. At the same time, the company applies pressure by putting outside producers on notice that certification is quickly becoming a requirement in all purchase contracts. BBP certification involves strict prohibitions on the use of chemicals (Paraquat, for example). The farmer will almost certainly have to make infrastructure changes such as providing buffer zones between waterways and banana rows, and planting native species to provide corridors for wildlife. Ultimately, the farmer will also have to be certified to SA 8000, which demands adherence to a range of ILO conventions, including freedom of association and prohibition of child labor (see Appendix A).

If the farmer sells to Dole, Del Monte, or Fyffes subsidiaries, he most likely will be required to set up an environmental management system conforming to the rules of ISO 14001, the international business standard that integrates environmental management systems with other management systems. He may opt to steer clear of the toxic agrichemicals Paraquat or Bravo, for example, but the certification does not require him to do so. If he sells to Noboa in Ecuador, no certifications are required. There the farmer is subject to Ecuador's labor law, which is indifferently enforced—although this may change as a result of a new government.

Pilot projects are currently under way to field test a set of social standards common to SA 8000, SAN, IFOAM (organic), and IFLO (Fair Trade). By far the majority of bananas in international trade under any of these four certifications are Chiquita's, certified by SAN as Better Banana.

The United Kingdom is by far the largest importer of organic bananas, and there is general agreement that this is because the big supermarkets there have decided to give them more shelf space. FAO interviews in Europe reveal that more than half of consumers have a low level of information about organic farming and organic labeling. Consumers are confused and not very well informed by marketers. There also exists some distrust in the reliability of organic labeling on imported products, except in Switzerland, where consumers are fairly well informed. As in Europe, American health concerns and the willingness of a segment of the population in a number of cities to move to organic produce have helped the organic banana market. The increasing presence of chains of organic supermarkets such as Wild Oats has also helped sales.

Some big banana farms have had success in converting from standard bananas to organic, a shift that worries small-farmer advocates. But the greatest obstacle to organic banana production is Black Sigatoka. Where the incidence of this fungus is low, sprays with organic fungicides may be sufficient. For both organic and standard banana producers, the development of a Sigatoka-resistant variety could be a real boon, since the treatment by aerial spraying is by far the most controversial health and worker issue in the industry.

At its meeting in Costa Rica in 2001, an FAO working group on bananas and tropical fruits found that both Fair Trade and organics face very similar goals in order to gain greater consumer acceptance outside their niche markets: "The 'common' challenge is the demand for fair-trade bananas to be organic and for organic bananas to be socially just." These niche rivals may need a fairly capacious tent in which to gather.

The Rainforest Alliance and Chiquita so far have been able to weather the rough seas of label hostility. They have survived and prospered as institutions in a transatlantic cultural and ideological contest with very high environmental and social stakes. Chiquita and the Alliance are essentially brand-driven organizations that face tough customers on both sides of the Atlantic. Although a small glimmer of consumer and wholesaler understanding of the potential of sustainable development branding seems to be visible, the many labels can only survive by passing severe tests of transparency and credibility. And that is true of ISO, SA, IFOAM, FLO, and the rest of the auditor's alphabet.

12

Unfinished Business

R inging down the curtain on this fast-moving unfinished drama is frustrating. Before we do, we want to tie together the loose ends of our story and show its linkages with the realities of this parlous era of terrorism, war, uncertain economies, global poverty, environmental deterioration, corporate scandals, and relentless anti-Americanism. Even if it were not sufficient to solve all the challenges, the moral and financial clout of a growing collection of certifiably responsible multinational companies would certainly be a welcome new ally in any unified drive toward sustainable development.

The state of play has three parts: loose ends, the big picture, and open questions.

The Loose Ends

With the banana-trade conflict over for the time being, in the spring of 2002 a tremendous bunch of bananas moved about in world trade (an annual rate of slightly more than 11,655,000 tons, according to FAO). Chiquita, still shipping its share, was a much different company. Here are some of the ports visited and volumes shipped in just the last several days that April: Four Chiquita banana refrigerator vessels, in a typical pattern, left the port of Limón in Costa Rica headed for points north: the *Chiquita Bremen* steered for Göteborg, Sweden, on the 22nd; the *Frances L.* headed toward Wilmington, Delaware, on the 25th; on the 28th, the *Chiquita Deutschland* left for Bremerhaven and Göteborg; and

---

also on the 28th, the *Lady Kokula,* with more than 20,000 boxes from Costa Rican farms, sailed for Antwerp. A week or so later, two Chiquita ships left Guatemala for Gulfport, Mississippi, and Miami, Florida, with over 300,000 boxes on board. Between April 26 and May 2, twenty-eight ships loaded with the yellow fruit headed north from Ecuador. Two of these carried Better Banana–certified, Chiquita-branded bananas from farms owned by Reybancorp and headed for Gdynia, Poland, and St. Petersburg, Russia. In the month of February 2002, a total of sixty-six vessels loaded with Chiquita bananas left the Philippines for ports in the Middle East, China, South Korea, Hong Kong, and Japan. One exporter alone (the Soriano group) shipped more than 900,000 boxes to Japan. Chiquita's owned farms in the Philippines have now been certified to Better Banana and the fruit is sold to Japanese supermarkets.

A hurricane struck again in the last weeks of April and into May. Driving rains hit the banana-producing provinces in the Atlantic coast regions of Costa Rica. More than five thousand people were evacuated. The Estrella Valley near Limón was especially hard hit. The banana industry suffered damage to infrastructure and several roads. Row upon row of banana plants were damaged or destroyed. An analyst estimated the quantity of boxes lost to export over the next month at 100,000 to 120,000. According to CORBANA, roughly 4,000 hectares of banana plantations were flooded. Some estimates were twice that high.[1]

Despite the ever-present role of the weather, Chiquita's new style was beginning to pay off, and some important organizations were looking at the company in a new light. On May 6 Chiquita was able to announce the good news that it had joined the Ethical Trading Initiative, along with the British company Fyffes. ETI is a unique alliance of companies, non-governmental organizations, and labor unions working together to advance business ethics, corporate responsibility, and human rights. ETI auditors had been on Chiquita's farms in Costa Rica the previous winter and saw the changes for themselves, after years of a distinctly cool relationship with the company. George Jaksch, now director, corporate responsibility and public affairs for Chiquita Fresh Group–Europe, was able to say, "We look forward to both learning from the organization and sharing our experience in the transparent implementation and monitoring of international labor standards."

Financially, things were looking up as well. On May 8 Chiquita's first

---

Chiquita after Reorganization, 2002

For 2001, Chiquita reported $155 million in earnings before interest, taxes, depreciation, and amortization, and before reorganization costs and other unusual items—an improvement of 7 percent over the previous year. Overall, the company reported a net loss of $119 million, including $78 million of accrued but unpaid interest later eliminated in debt restructuring; $34 million of reorganization costs; and $28 million of other unusual charges primarily associated with farm closures, and a third-quarter strike and related labor issues at the Armuelles (Panama) division.

Chiquita hosted a meeting of analysts on September 24, to discuss the results of the months-long "strategic review" undertaken by the new management. The company outlined various profit improvement initiatives, including overhead reduction, centralized purchasing, and productivity—improvements in fresh produce and processed vegetables, among others. It targeted a minimum of $255 million in earnings before the usual deductions for interest, taxes, and the like.

Management also laid out plans to improve its cash flow and said it was looking for $100 million to $150 million in asset sales over the next two years. (Fifty million dollars had accrued from the sale of five ships.) Another goal was to reduce working capital by $100 million over the next two to three years.

---

annual report after emerging from bankruptcy in March showed a first-quarter net loss but increased banana demand and higher prices (see box). Restored import levels under the new banana regime enabled Chiquita to recapture some of its lost market, but at a painful price to the small farmers in former colonies in the Caribbean (and, therefore, to Chiquita's rival Fyffes and to the Fair Trade bananas that had been protected under the previous system). In 2006 the EU plans to move to a tariff-only system, the ramifications of which are uncertain.

In November 2002 the company published its second corporate responsibility report covering the previous year. The report was divided into the "triple bottom line" (environmental, social, and financial performance), with an added section describing progress on the company's various "core value" targets (see Appendix C).

Coincident with Steven Warshaw's elevation to president and chief operating officer in 1997, Keith Lindner had ceased having any operating role at Chiquita and moved into a copresidency with his brothers in the Lindner family parent insurance company, American Financial Group. That left Warshaw with the responsibility of dealing with all the problems that crashed down on the company in 1998. But it also gave him a freer hand to distance the company from its unsavory past. With all the trouble in 1998, the Lindner family's role in the operations of the company began to diminish. Perhaps Warshaw's decision to embark on the corporate responsibility drive, including the outreach to former antagonists, together with the fading Lindner presence, had the effect of softening the rock-ribbed, Midwest Republican personality of the company—a personality that had conflicted sharply with the more socially conscious European activists.

In May 2001 Carl Lindner announced his retirement as board chairman. He and his family reduced their ownership stake from more than 33 percent to less than 10 percent. In his statement Lindner said that at the age of 83 he wanted to spend more time with his family and attend to the affairs of American Financial Group. Warshaw, now chief executive officer, had the thankless job of guiding the company through a complex bankruptcy and reorganization. Robert Dunn, president of Business for Social Responsibility, says that Jeff Zalla's calm hand on the reins of the corporate responsibility program was a critical factor in keeping the company moving forward during this treacherous retrenchment. Then, in March 2002, Chiquita announced that Warshaw had resigned and been replaced by a former Booz Allen Hamilton vice chairman, Cyrus F. Freidheim Jr. Warshaw reportedly received a $4 million severance package. A new board was appointed, but among top managers both Zalla and Bob Kistinger survived. To most outsiders, the company's balance sheet looked promising. One analyst even issued a "Strong Buy" rating on Chiquita shares. The company's stock began to appear on the buy lists of a few social-investing groups. All in all, it was a remarkable turnaround for a company that had been literally shunned by buy-side brokers for several years.

The company was even beginning to get grudging accolades from European antagonists. Banana Link's Alistair Smith reported to us in an e-mail that his organization's work had shifted to ensuring implementation "on the ground" of the June 21, 2001, COLSIBA-IUF-Chiquita labor agreement. He said that Banana Link regarded this responsibility as much more important than the certification initiative by the company itself. Smith added that Chiquita's corporate responsibility report was "quite impressive and surprisingly frank, especially in the introduction, compared to the abysmal historical record of the company. They are clearly making a genuine corporate effort at the top, but they have serious problems within certain countries with the antiunion culture/ attitudes of local management, especially in Costa Rica and Guatemala." He concluded by noting that Banana Link was now turning its attention to treatment of union workers on Dole and Noboa farms in Ecuador. In addition, Banana Link had been pushing a Fair Trade initiative in St. Vincent and St. Lucia in the Windward Islands. Growers there were exporting four thousand boxes of Fair Trade bananas a week in 2002, earning a "social premium" for the farmers of more than $390,000 a year. Then tropical storm Lily struck in late September, devastating many farms on the islands. In a 2003 note from Smith, he relates that "Exports in the first quarter of 2003 shrank from over 28,000 tons for the four islands put together to less than 18,000. Drought in Dominica and St. Vincent accounted for much of this fall. This means that Fair Trade exports from farmers' groups organized by the Windward Islands Farmers' Association now account for more than 15 percent of total exports. Despite its initial reluctance to accept the strategic value of Fair Trade, the industry is starting to see that it may play a vital role in the survival of the islands' most important export."[2]

Old Enemies and New Friends

Chiquita continued to have its critics, from disgruntled former employees and cynical day traders on the Yahoo finance message board to *Fortune* magazine. In a November 2001 article, Nicholas Stein wrote that

Chiquita's story goes beyond unfair quotas and bad luck. He criticized the company for making risky bets and poor decisions that angered customers and disappointed Wall Street. Stein accused the company of gross management blunders in its reaction to the shrunken European market.

Others have criticized Chiquita's single-minded reliance on lawyers and connections in Washington to fight the European regime, instead of investing in farms in Africa and the Caribbean so that they would receive favorable treatment under the quota and license system. Few employees could have been happy at the value of the stock in their pensions when the company declared bankruptcy (although the company states in its 2001 corporate responsibility report that employee turnover during the troubled times was not above average). *Fortune*'s Stein wrote that the end of the trade war had left the company in a relatively favorable state, and that "the company has owned up to its spotty environmental record and is taking steps to improve it." But he added, "If it wants to emerge from Chapter 11 with a chance to succeed, the company similarly will need to acknowledge its corporate missteps." In fact, in its candid and comprehensive corporate responsibility reporting, Chiquita has done just that.

We authors were not the only ones to think so. Late in 2002, SustainAbility (U.K.), an international consultancy specializing in business strategy and sustainable development, ranked Chiquita's 2001 corporate responsibility report (covering 2000) as number one in the food industry and number three among all U.S.-based companies. In announcing the prize, SustainAbility editors wrote of Chiquita's report, "Its coverage includes some courageous accounts of weaknesses and failures, some of the most candid and honest reporting we have seen to date." SustainAbility's survey had identified and analyzed what were, in its view, the fifty top reports from around the world. Robert Dunn of BSR said in our November 2002 interview: "For us the partnership with Chiquita Brands is one of our proudest moments because it affirms the purpose of this organization, namely, it's possible to influence policies and practice and prove that companies can still be profitable in a way that benefits society. We have made a contribution to this journey and we're very proud."

At this point, Chiquita no longer has to rely on the enthusiasm of a few believers to praise the positive changes the company has already made on the environmental side, and its promising social and core value commitments. With respect to the benchmarking flowing from the company's labor goals, German Zepeda, president and coordinator of the banana and agricultural unions of Honduras, is quoted in Chiquita's 2001 corporate responsibility report: "These assessments were conducted, and this report written, by a more responsible Chiquita—a company that has been steadily transforming itself over the last several years." Stephen Coats, executive director of U.S./LEAP, a vocal champion of Central American agriculture workers and an early critic of Chiquita, says in the report: "Chiquita continues to impress. The transparency of its corporate responsibility reporting and the use of highly respected independent observers . . . combined with some real progress on the ground, is a track record which is unmatched in our work in Latin America. While the road is long, Chiquita has traveled far in a few short years." To persistent skeptics, Robert Dunn cited Goethe: "We see only what we look for and look for only what we know."

The reality is that, commitments or not, unions and impatient stockholders have their own demands. Early on, Freidheim was scolded by stockholders for withholding future short-term results in favor of the long-term stance he perhaps understandably adopted in a give-and-take with key shareholders. Under pressure to get rid of unprofitable assets, the company was negotiating the sell-off of Armuelles, its Pacific operations in Panama, after losing some $90 million in the previous five years. The company wanted to sell the farms to the workers, with the proviso that Chiquita have the sole right to continue to buy the fruit from a new worker-owned cooperative that would be formed in the deal. At this point the workers rejected the offer, because they wanted the right to sell to other interested operators. Chiquita cut its subsidies to the division and still hoped for a negotiated deal. Perennial strikes by well-organized laborers demanding higher pay as well as high irrigation costs because of lower rainfall than on the Atlantic plantations had driven the impasse in Panama.

After a protracted and difficult negotiation, in April 2003 Chiquita signed an agreement with union leaders and the government of Panama to sell the Puerto Armuelles unit to a cooperative led by members of the Armuelles banana workers' union. Under the agreement the company was able to overcome the previous sticking point, and the cooperative/

union hybrid agreed to sell fruit solely to Chiquita at market prices for a ten-year period. Chiquita says the sale means that 2003 will be the last year it will report on the environmental and social performance of the twelve farms involved. But the company insists that it will continue to work with the new owners to maintain improved performance in these areas, as it is doing with all of the other independent producers.

To reduce debt further, Chiquita sold its vegetable canning subsidiary to Seneca Foods Corporation for $110 million in cash and 968,000 shares of Seneca preferred stock. Seneca also assumed the subsidiary's debt of some $88 million. That spring the company bought Atlanta AG, a leading German distributor of Chiquita products in Germany and Austria that has been Chiquita's largest customer in Europe for many years. By exchanging loans for Atlanta's underlying interests, Chiquita was able to acquire ownership in a virtually cashless transaction. This acquisition increased Chiquita's consolidated revenue by about $1.1 billion.

In the same period Chiquita reported first-quarter net income of $25 million; in the year-ago quarter the company had reported a net loss of $398 million. North American base prices rose 3 percent from the prior year. The volume of bananas sold in North America fell 5 percent from the year-ago quarter because of the flood in Costa Rica in December and the company's decision to sell less fruit on the volatile spot market. Chairman Cyrus Freidheim was quoted as saying, "Strategically and financially we are in a much better place than we were a year ago." By the middle of 2003, new stocks reissued after the bankruptcy finally broke through a persistent fifteen-dollar ceiling and a number of anonymous commentators on the Yahoo message board were issuing "strong buy" recommendations.

It was during this period that Jeff Zalla, who had been vice president and corporate responsibility and communications was promoted to vice president, treasurer and corporate responsibility, making him responsible for treasury operations, banking, credit, cash management, and financial analysis—heavy lifting for a man not yet 40 years old. In continuing his oversight of the company's corporate responsibility program, Zalla reports to the audit committee of the board of directors. This move places corporate responsibility at the heart of company financial management and in a strong position to influence Wall Street thinking on the practical value of corporate responsibility.

At the same time, Chiquita earned the first ever CERES-ACCA award for outstanding sustainability reporting for its 2001 corporate responsibility report. The Coalition for Environmentally Responsible Economies (CERES) and the Association of Chartered Certified Accountants (ACCA) presented the award at a conference attended by companies, NGOs, investors, and financial analysts. CERES, since its founding in 1989, has persuaded dozens of companies to endorse its sustainable development principles and co-founded the Global Reporting Initiative (GRI), which prepares guidelines for reporting and has attracted growing interest from many progressive companies. Chiquita shared first prize with Ben and Jerry's Homemade Holdings (now owned by Unilever). The judges said that Ben and Jerry's continues to set an example for smaller companies in reporting qualitative and quantitative information. Moreover, the company was the first brand-name firm to sign on to the CERES principles and has been producing reports based on the CERES format for more than ten years. Since Chiquita does not produce its report in conformance with the GRI format, the judges simply were unable to decide between two such different reports. It was at this conference that Alice Tepper Marlin of SAI and Jeff Zalla spoke in favor of enlarging the scope of the GRI guidelines to include reporting of company independent certifications.

Our purpose has not been to judge Chiquita's overall business record or to second-guess its various business decisions. The record will ultimately speak for itself. In the areas we have investigated, there can be no question about the environmental progress the company has already made. Its public commitments to labor rights and its internal dedication to a set of core values are very promising. In fact, virtually all of the company's former critics agree that through the BBP and SAI certifications, frank and transparent reporting, and the unusual pacts with two important union groups, Chiquita has set a new standard for the industry—in fact, for any industry.

Moving Ahead and Getting Together

The Rainforest Alliance also had a busy year in 2002. With Dan Katz now chairman of the board and Tensie Whelan executive director since 2000, the Alliance had become an internationally recognized NGO. Still relatively small, it had a clear niche in environmental certification, ambitious plans for the future, and a budget of $6 million—shooting for $8 million. Among other tasks, Whelan's calendar that May included planning for a major office expansion nearby in downtown Manhattan at the end of the year and a meeting with the head of the sustainable agriculture initiative for Nestlé. She also met with staff from the environmental ministry of Mexico to chart future joint projects. A long-overdue planning initiative for the agricultural division was launched that month.

Whalen, her staff, and the board continue to grapple with the problem that has lately plagued the accounting and consulting industries and led to the present state of distrust in that industry. As independent certifiers, how can Rainforest Alliance auditors and their partners perform and be paid for their various professional services to clients and at the same time steer clear of any perception of conflict of interest? In a perfect world, certification as a program of these nonprofit organizations under SAN would be self-funding. The Rainforest Alliance continues to seek that holy grail of most NGOs: unrestricted grants from foundations and individuals. But foundation priorities can be fickle and often change in response to new and trendier issues. The organizations carrying out the certification audits must be wary of the charge that they cannot be objective if they work too closely with the client they are certifying. Although skeptics remain, the integrity of the Rainforest Alliance is now apparent to most. Meanwhile, its SmartWood program continues to expand, with certifications in North America and Africa, and SAN is certifying flowers, ferns, citrus, and cacao throughout Latin America. By midyear 2003 Whelan ramped up an already busy schedule and began to consolidate contacts in Europe.

Early in 2002 the beginnings of a rapprochement among the chief international agricultural labeling schemes were becoming apparent. Under a series of pilot programs in a number of countries, a new umbrella group calling itself Social Accountability for Sustainable Agriculture (SASA) was formed.[3] Its mission was to look into ways to integrate audits so that farmers would be able to attest to both social and environmental responsibility under a single audit. This was indeed a major

---

About SASA

SASA is a research project of iSEAL, an umbrella group of accreditation bodies and labelers that includes organic, Fair Trade, the Forest Stewardship Council, Social Accountability International, and the Rainforest Alliance's Sustainable Agriculture Network. Members of iSEAL first came together informally in 1999 in Oaxaca, Mexico, hoping to find ways to ensure that their activities were in line with international norms and not perceived as nontariff barriers to trade. In short, the constituents wanted to work with one another to gain political recognition. The group is taking the necessary steps to achieve official ISO accreditation and conformance with WTO prescriptions. Mindful of the organic standard set by the U.S. government, the group is also determined to monitor carefully any other government interest in the private labeling area.

The leaders of SASA initiated joint pilot projects beginning in June 2002 in Costa Rica, with SAN leading the group on bananas. Wille says that the SASA pilots are going well. "The groups have made terrific progress from a history of mudslinging to mutual respect and understanding of each other's programs. I think we've done enough pilots, and now we need to find the time to sit down and work out a common language." In essence, the members hope to gain a common understanding of what a "social agenda for sustainable agriculture would entail and an agreed framework for working in that direction."

---

step to reduce consumer confusion about labels and to seek common ground among the players. If successful, Chris Wille's initial fears of a "Darwinian shakeout" of labels could turn out to be unfounded. Instead, the single-audit approach presages an amalgamation among the labels whereby each can retain its unique mission and public image while combining forces to make certification easier for farmers.

FAO professionals had a role in bringing the principal players together in earlier meetings where some of the key labelers became better acquainted. Leaders in the commodities division had promoted the idea at the 1999 meeting of the FAO banana group in Australia. In Rome a year later, there appeared to be a real breakthrough in relations, as organic, Fair Trade, and Alliance people were able to find at least some common ground among labeling principles and practices.

By early 2002 the Sustainable Agricultural Network had nine partnership organizations in Belize, Brazil, Colombia, Costa Rica, Denmark, Ecuador, El Salvador, Guatemala, and Mexico. (A group in Honduras subsequently joined.) All had fully trained certifiers and were active on farms of a number of crops throughout the Latin American tropical zone. By 2003 they had certified more than 46,000 hectares of banana farms to the Better Banana attestation, nearly 8,000 hectares of coffee, over 7,000 hectares of citrus, and about 4,700 hectares of cacao.

Also in that year, the Alliance's agricultural group sent a representative to a workshop on "financing green goods and services," convened by the North American Commission on Environmental Cooperation (CEC), the environmental arm of NAFTA. The CEC template has been promoted by some as a practical way to accommodate environmental considerations in multilateral trade agreements, still unsettled in the WTO. In addition to representatives from the Canadian, Mexican, and U.S. governments, people from the World Bank, Starbucks, and Green Mountain coffee attended the workshop. The CEC promotes ways to use market forces to foster environmental protection, and the Rainforest Alliance has agreed to help build the business case.

The Sustainable Agriculture Network has an ambitious plan: building from a current base of 63,000 hectares, the network will increase the amount of acreage across all crops to approximately 125,000 hectares by 2006. The growth rate for bananas will be nearly flat; most of the growth will be in coffee and cacao. And it will come through large new projects, such as a sustainable coffee initiative and contracts with major coffee roasters and supermarkets. The network plans to triple the number of farms enrolled in the banana certification program from about 330 to 1,000. But its work with bananas, while not growing much, will become more intense. Chris Wille says, "Chiquita wants SAN to challenge them and guide them in environmental performance." He asserts that he is determined to provide progressive and innovative audits in even the most difficult banana farms.

Food Safety Fears

Another hurdle for the Alliance and its various certification partners has arisen in Europe. Food retailers there created EUREP-GAP, a new certification scheme covering all fresh produce. The program started in 1997 as an initiative of retailers belonging to the Euro-Retailer Produce Working Group (EUREP). The aim was to agree on standards and procedures for development of appropriate agricultural practice. Essentially, it involves an accreditation process through which certification bodies are empowered to grant a certification without which exporters such as Chiquita would be barred from selling to the big retailers in Europe.[4]

Although initially triggered by European retailers' perception that consumers needed reassurance about food safety on the heels of mad cow disease (BSE) and the introduction of genetically modified foods, it is hard not to suspect a certain defensive cultural cast to EUREP-GAP.[5] Some retailers are saying that all their suppliers must be accredited by EUREP-GAP by 2004. Others do not have a deadline, but may eventually question why preferred suppliers are *not* EUREP-GAP certified and perhaps review their decision to do business with them. Chiquita and the other dollar-banana companies might well have European certifiers roaming their farms in the tropics, checklists at the ready, grading farm managers to environmental standards lower than Rainforest Alliance certifications (except, as Chris Wille notes, for certain protocols on food safety).

To cope with this new impediment, the Sustainable Agriculture Network decided to become accredited to EUREP-GAP. As a certifier of Chiquita's farms and the farms of others who sell into the European market, it could hardly do otherwise. So in mid-September 2002 Wille, by that time undeniably a veteran in coping with these matters, addressed a technical workshop in Madrid convened by EUREP-GAP on the "issues that environmentalists find interesting, innovative, or problematic within the EUREP-GAP standard."

He expects that all banana producers will sooner or later obtain EUREP-GAP and/or BBP accreditation, and then "snap on auxiliary certifications," especially SA 8000. All agribusiness companies will eventually adopt ISO, he thinks, as a cost of doing business. It is only rational to organize along internationally accepted quality, service, and environmental guidelines. Wille further asserts that "social issues eclipsed environmental issues for the past few years, but environment is coming back strong. We eventually expect to be able to offer triple certification: BBP, SA 8000, and EUREP-GAP."

Wille says that the EUREP-GAP standards touch most social issues but not conservation issues. "The standards are good on chemical use but not as detailed. It is a pass-fail program, so there is no incentive for continuous improvement. It is professionally managed, constantly reviewed, and republished every several years. EUREP-GAP has not worked hard to be open to anyone wanting to participate, including NGOs. The auditors don't have the kind of local knowledge that BBP auditors have. It is second-party auditing. But since many European supermarkets, especially the large ones like Ahold, demand EUREP-GAP certification, I have urged NGOs to get on the inside and wrestle with them."

The new competitive landscape, according to Wille, is not among the NGO-backed certification programs like IFOAM and Fair Trade, and SA 8000; the real competition is between the SASA members and industry-supported certifications. When BBP is finally accredited by EUREP-GAP, Chiquita will not have to do any more to meet European standards. It is unfair that Chiquita has to pay for both EUREP-GAP and BBP.

Banana Link's Alistair Smith told us in an e-mail:

> Clearly the European supermarkets are throwing their weight around by trying to combine quality and environmental and social standards into their buying requirements, without consultation as usual, without remuneration for the efforts of suppliers as usual. Indeed, producers have to pay dearly for this; the big ones can afford it, but the others are likely to be squeezed out even faster than they are being squeezed by supermarket pressure on prices. EUREP-GAP and their new "three in one" standards are probably more of a threat immediately to the Rainforest Alliance, as there will be much overlap in requirements, making their efforts marginal. All in all, it is good that we have supermarkets starting to recognize the need for supply chains to meet a range of ethical standards, but they need to pay a fair price to allow their suppliers to achieve what they want; otherwise we

have further concentration of power at the top end of the chain with growing unemployment and poverty at the production end.

The Big Picture

The Seattle experience showed that NGOs—from local to international— have become increasingly powerful, and that intergovernmental institutions have been slow to develop a meaningful way to deal with them. Tear gas and water cannons are clearly not the answer. And as international diplomats have discovered, many of the street hippies of the 1970s have morphed into sophisticated three-piece-suit advocates, easily their match in sophisticated debate—but activists can morph back into street insurgents at the click of a mouse.

After the Seattle debacle, economist Yash Tandon, editor of the *SEAT-INI Bulletin,* a journal promoting Third World views, put the future of clashing agendas presciently in an e-mail to us in which he said he was convinced of two things. One is that the rich countries have no choice but to keep on manipulating the system and making decisions undemocratically. He feels that an open democratic system is not in their favor. His second conviction is that the NGO–Civil Society movements, with all their contradictions, will become stronger and will use the Internet to make life difficult for WTO, IMF, World Bank, multinational companies, and U.S.-EU governments. "And they will be right to do so," he concludes.

Multinational corporations, for their part, have no choice but to engage globally. Most no longer boast of the national origins of their products; instead, they rely increasingly on global brand names and polished corporate images, making them increasingly vulnerable to the rancor and ridicule of NGO campaigns. The walls between campaigners and corporations were beginning to crumble in the 1990s. Considerable confusion arose in the corporate ranks, for instance, when BP and Shell included human rights and sustainable development issues in their corporate annual reports.

In 1999, just after the street uprisings in Seattle, Georg Kell and John Ruggie, senior staff advisors to UN Secretary General Kofi Annan, presented a paper to an international conference at York University in Toronto. "Global Markets and Social Legitimacy: The Case of the 'Global

Compact'" provided a generally persuasive intellectual framework for Annan's controversial initiative presented to the Davos World Economic Forum the previous January. Annan had challenged the private-sector movers and shakers to help the UN implement "universal values" in the areas of human rights, environment, and labor through this global compact. While 850 large companies around the world have signed on to the compact, most big American brand-name companies have so far declined. However, as of 2003 forty-four U.S. corporations have joined, including Cisco Systems, Dupont, Hewlett-Packard, Nike, and Pfizer. In the U.K., out of twenty-two, BP, Rio Tinto, Royal Dutch/Shell Group, and Unilever are members.

The International Chamber of Commerce adopted a statement in July 1999 arguing for a stronger UN and promising to help bring companies into the fold. The ICC realizes that a stronger UN in the areas of human rights, labor, and the environment is the best strategic path to secure open markets. It remains to be seen whether the corporate side can muster enough "moral first movers" to establish industry-wide commitment to these goals. The potential contamination of the global public policy process by the close involvement of the private sector worries a number of NGOs. Others, while recognizing the enhanced possibilities for corporate probity, have reservations about the motivations of some companies and the mechanisms for enforcing commitments to the global compact. Amnesty International, for example, and other human rights groups have called for an "ombudsperson" and mandatory, regular reporting by the firms of actions they have undertaken to demonstrate their practical commitment to the goals of the compact.

The reality is that more than three hundred project partnerships were submitted to the 2002 World Summit on Sustainable Development in Johannesburg, South Africa—many of them involving private companies. It suggests that "business as usual" and the slow progress by governments on sustainable development may face competition from the private sector–NGO agreements as the model of global problem solving.

These partnerships and similar joint ventures, sometimes between long-established adversaries and outside the purview of governments, may be a breakthrough vision of environmental problem solving. While distrust of corporate motives and of government obfuscation was the crux of the discontent in Seattle in 1999, a counter trend is surfacing that presents a much different view of the potential role of corporations in solving global problems.

Jo'burg Jazz

The EUREP-GAP workshop on food safety issues was convened in Madrid only days after the conclusion of the ten-day summit in Johannesburg. The agenda of this megaconference—attended by more than nine thousand delegates, eight thousand registered NGOs and industry delegates, and four thousand media people—covered this topic and a wide range of issues encompassing the disparate concerns of over 180 nations. The actions adopted set some target dates for change. Governments did agree to halve by 2015 the number of people who lack access to clean water or proper sanitation, and they established goals intended to protect or restore ecosystems; for example, the restoration of fisheries by 2015 and the reversal of biodiversity loss by 2010. Agriculture and food security would remain a high international priority.

As far as partnerships are concerned, Chiquita and the Rainforest Alliance were very early movers. At the Madrid meeting, Wille was in dialogue with business people from the vantage point of a program that had already worked for more than a decade; in South Africa, partnerships and other combinations between campaigners, corporations, and governments were just being acknowledged as part of the global system—ten years after the Rio Summit and thirty years after the Stockholm conference, where these collaborations barely seemed plausible.

Jonathan Lash, president of the World Resources Institute (a Washington, D.C., research institution) was outspokenly disappointed about the lack of progress by governments at the Johannesburg summit, but was enthusiastic about the new partnerships fostered there. He said, "The summit will be remembered not for the treaties, the commitments, or the declarations it produced. Instead it will be seen as the first stirrings of a new way of governing the global commons—the beginnings of a shift from the stiff formal waltz of traditional diplomacy to the jazzier dance of improvisational solution-oriented partnerships that may include nongovernment organizations, willing governments, and other stakeholders." Intentionally or not, Lash was echoing one of three

*"Betrayed by Government." NGOs demonstrate during the Johannesburg
Conference of 2002. Courtesy of IISD / Leila Mead.*

sustainable development alternative scenarios offered in 1997 by the
World Business Council for Sustainable Development, the Geneva based
organization of some hundred large multinational companies. Council
president Bjorn Stigson described one of the scenarios as Jazz.[6]

In the world of Jazz, disparate interests join ad hoc alliances to prag-
matically address social and environmental problems. The theme of the
scenario is dynamic reciprocity. It involves a coming together of social
and technological innovation, trying out different approaches, quick
adaptation, voluntary combinations, and a sturdy and constantly shift-
ing set of global markets. What makes the rapid learning and subsequent innovation possible in Jazz is transparency. That is to say, per-
vasive availability of information about how products are made and
grown; the sources of manufacturing and farming inputs; the corporate
financial, environmental, and social performance; the government negotiating processes; and almost anything else that concerned publics
want to know.

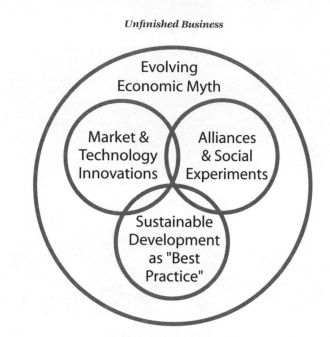

"Jazz" scenario from the report by the World Business Council for Sustainable Development. SOURCE: "Exploring Sustainable Development: WBCSD Global Scenarios, 2000–2050," summary brochure (Geneva, 1997).

While these scenarios are not new and while not all audiences endorsed them, the combinations that surfaced from Johannesburg bear an uncanny resemblance to the Jazz scenario, as does the story of Chiquita and the Rainforest Alliance.

NGOs at the Table

In reflecting on the outcomes of the World Summit on Sustainable Development, Noel J. Brown, former director of the North American Regional Office of UNEP, newly returned from Johannesburg, said that perhaps the critics of the summit did not really understand the complexity of what was being tackled there. The consensus-building process behind international environmental treaties is convoluted and cumbersome. Governments needed fully twenty years to negotiate the Law of the Sea treaty, for example. "The problem is that all government stakeholders don't have the same level of interest, but they are all sitting at the same table." Johannesburg was an example, Brown said, of a new international dynamic. Government agenda setters had to consider a smorgasbord of wishes and pledges from the Rio Summit; world trade and finance issues; and aspirations for peace, ending poverty, protecting the environment, and human rights, among other goals set down in the UN Millennium Declaration. Moreover, outcomes depended on negotiations among new combinations of governments, activists, and companies.

What Brown calls influentials—both individuals and groups—are today very important in making things happen at the UN. Unlike governments, they can focus on single issues of great urgency to them. That is why the partnerships negotiated at Johannesburg have real promise. "We've come a long way from Stockholm and even from Rio," Brown said. At those summits NGOs were guests of grudging governments, but now they share responsibility with the hosts. "Protest is not policy," he concluded; "NGOs came of age in Johannesburg, but now comes the hard part."

In fact, the NGO movement has grown by almost 30 percent worldwide since 1990. These activist organizations are heavily concentrated in Scandinavia, the Benelux countries, the U.K., Germany, Austria, and Switzerland. According to a 2001 study by the London School of Economics, fully 33 percent of the membership of these international NGOs is in Western Europe, while a meager 4 percent is in North America. Although these are extremely diverse groups, the study classifies their political positions on globalization into four categories: single-minded supporters; "rejectionists," including right-wing nationalists; reformers, who accept globalization but want to civilize the process (the largest group); and so-called alternatives, who seek a lifestyle without interference from governments, companies, or international institutions. In recent years these groups have mounted significant alternative summits parallel with those convened by the larger institutions. Still, partnerships are unlikely to supplant protests as the globalization debates heat up.[7]

Market Realities

A successful linkage of the agricultural certification process into the global marketplace requires confronting some realities. An organization that works with markets to achieve social transformation must inevitably face the nature of the markets themselves. All the complexity, competition, subtlety, uncertainty, bluster, cheating, and government measures to deal with them persistently come into play. Although business history teaches that players will go to great lengths to get around it, the law of supply and demand rules. The truth has been that labels do not always persuade the middlemen in the banana industry.

One distributor in the U.S. Northeast, who prefers to remain anonymous, explained that bananas are a big challenge at retail. The consumer wants them "somewhat ripe." But bananas from a given company often vary considerably in quality from month to month. During certain times of the year bananas from one country are more marketable than bananas from another. "When Costa Rican fruit is not the best, Ecuadorian fruit often is." But bananas from Ecuador, from farms owned by or selling to Noboa and Dole, are produced by workers who labor under substandard conditions, allowing the fruit to be sold cheaply in the northern markets. Good-looking bananas at a bargain price are practically irresistible to most distributors and to their customers, the supermarkets. This quality-and-price equation is what drives the oft-cited "race to the bottom."

Moreover, certification must learn to adapt to national differences. We have reported that bananas grown on Ecuadorian farms selling to, or owned by, Reybancorp are largely certified by Better Banana and sold under the Chiquita label. Wage conditions are problematic even so, but a concerned buyer needs to know that workers on these certified farms, despite the fact that most are not protected by union agreements or Ecuadorian law so far, increasingly benefit from certification standards that include values enshrined in the ILO conventions. That is because those standards are now incorporated in the Better Banana certification and will, according to Chiquita's commitments, be audited under SA 8000 standards as well. The workers also enjoy a range of nonwage benefits, such as dental care. All of this, of course, costs money that should come back to the company somewhere in the banana chain. Still, without significant consumer pressure, retailers are not likely to budge far on price, as bananas are the most profitable items in the produce department and account for 10 percent of all sales. Moreover, a demonstrably responsible banana company like Chiquita, if it consistently

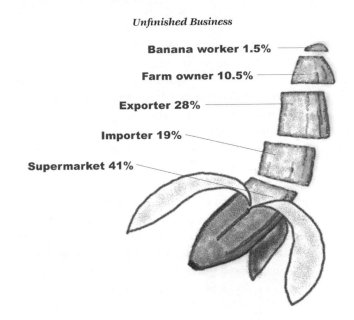

Banana worker 1.5%

Farm owner 10.5%

Exporter 28%

Importer 19%

Supermarket 41%

Who gets what from bananas bought in a UK supermarket? A portrait of the "race to the bottom" on bananas with a Bonita (Noboa) label grown in Ecuador. SOURCE: Banana Link, U.K., and New Internationalist, www.newint.org.

delivers high-quality fruit and is responsive to the other demands of supermarket chains, is in a position to affect the reputation of the supermarket.

The big companies strive for long-term contracts with supermarkets, but experience has taught them that it is difficult to deliver high quality consistently. "No one company can have good bananas all year round. Age of the field and weather are critical," the distributor said. "Basically it comes down to the classic discussion: the seller wants the highest price he can get, and the buyer wants the highest quality at the lowest price. The buyer at the pier assumes that all bananas are the same. He may take some from each company."

Yet there can be exceptions, and perhaps change is in the wind. One distributor we spoke to saw his role as educating the retailer. When there were news reports of worker exploitation on strawberry farms in California, he went to check it out. He didn't find any, he commented, but if he had learned that farms were using child labor or otherwise abusing workers or the environment, he would have seen it as his duty to steer clear of them and explain why to the retailer. It is not an impossible stretch to believe that a banana brand can come to mean more than color and curve. Organic products are finding more and more shelf space and many stores are selling Fair Trade and organic coffee. What is required with bananas is greater attention from the certifiers to get the message across to retailers and customers in the United States, in much the same way that Fair Trade has managed to communicate in the U.K. and elsewhere in Europe—and increasingly in North America as well, especially in Canada.

Supermarket consolidation is another issue. To land a big long-term contract with Wal-Mart, for example, would be a huge bonanza to any company. Marketers of fresh fruit, and of almost every other product sold in supermarkets, have a more difficult job than formerly. Before, they could count on at least a few of the many independent stores to be responsive to price promotions or rebates or quality arguments, and companies were able to tailor their marketing pitches. Now they face a single buyer for thousands of tons of their product: it changes the whole equation. The development of the EUREP-GAP standard is a keen re-minder of the power of these behemoths. Chiquita's Bob Kistinger says: "Wal-Mart buys from us. We have a good relationship with Wal-Mart and we have a disproportionate market share with them. Wal-Mart is a very good customer. We were named its worldwide environmental supplier of the year for 2000."

Supermarket complacency may not last, however. Sister Ruth Kuhn, the early critic of Chiquita and activist member of the Sisters of Charity in Cincinnati, has praised the company for its reform efforts. She has also said that a number of social investors, including the Interfaith Cen-ter on Corporate Responsibility with its 275 faith-based institutional in-vestor members, are now working aggressively with Wal-Mart and other supermarket chains on several issues including sweatshops. "At first, Wal-Mart was very closed and unavailable to us, but now they have been willing to meet with shareholders and to talk with us regarding our re-quests for a code of conduct, particularly in their work with contract suppliers."

Bananas are selling well again in European supermarkets, but the sensitivity about genetic modification may cause future complications. In theory, bananas can be genetically altered toward greater resistance to fungal diseases such as Black Sigatoka. Although a successful outcome would all but eliminate the controversial and expensive aerial spraying of fungicides, in the current climate it is unlikely that customers would buy genetically engineered bananas even if, at some level, they might value the curtailed spraying. Clyde Prestowitz, president of the Economic Strategy Institute in the United States and former trade hawk in the Reagan administration, commented in an op ed piece in the *New York Times* on January 31, 2003, that he believes the ban the European Union has placed on genetically modified food imported from the States is almost certainly a violation of World Trade Organization rules. He cautioned U.S. companies and trade officials to back off any pressure. His experience in selling to Europe taught him that the pressure will almost certainly cause a backlash. As he puts it, "Food is to European culture what free speech is to American culture."

But the United States did not heed his advice. In May 2003 the U.S. trade representative filed a case against the EU with the World Trade Organization. In response, the European Commission called it "misguided, unnecessary, legally unwarranted, economically unfounded and politically unhelpful." In fact, the European position had been that it had not banned GM crops permanently, but had adopted a moratorium while member states sorted out the applicability of the so-called precautionary principle in this case. This "look before you leap" tenet is what really divides the United States and the European Union The former fiercely advocates the use of hard science to support restrictions on trade, while the Europeans equally fervently believe that in instances where the science is uncertain it is better to be safe than sorry.

Midyear 2003, the EU announced new rules which, if accepted, would bring an end to the moratorium. The proposed rules on labeling and traceability of genetically modified organisms requires products— like corn flakes, for example—containing more than 0.9 percent of GM material to be labeled as such in a fashion as to carefully trace GM seeds from "farm to fork." The United States immediately voiced strenuous objections, insisting that the new rules would be economically ruinous to farmers and industrial farming companies who have been working with genetically modified seeds for some time. They also called the rules unworkable in practice and scientifically unjustified. Europeans, on the other hand, described the rules as undergirding a robust safe-guard system.

Argentina, one of the world's biggest grain exporters, also took strong exception to the new labeling and tracing system, saying in effect that setting up such a procedure was literally impossible and the costs—running to several hundred million dollars—were prohibitive.

Chiquita, which announced early that it was carefully researching the possibility of switching to a genetically modified banana resistant to the most persistent fungal disease, Black Sigatoka, has not reacted specifically, but said at the time that it would not proceed if consumers continued to have reservations.

As the EU and United States lock horns on agricultural tariff issues at the start of the WTO Doha round, it will be a renewed challenge to the organization's credibility. Meanwhile, this brutal trade impasse seems to be developing in a culture war already exacerbated by heavy doubts in Europe about the U.S. government's armed overthrow of the Iraqi government. A television account said it all: Intense coverage of shoppers in European supermarkets shows them closely scrutiniz-ing product labels—further testament to the new power of certified products.

For the independent activist certifier, too, the marketplace can be tough. Pressures to cave in to company priorities are often fierce. Trans-parency can be compromised by the exigencies of survival. Moreover, our language has no real handle yet on the novel relationship between certifier and company. While there is no confusion about the Alliance's role as auditor, at various times the organization has described itself as partner, other times as parole officer. Sometimes it describes Chiquita as "client." While clear to those in the game, this ambiguity makes the certifier vulnerable to critics.

Early on, one academic said that Better Banana certification would never be credible until a farmer understood that decertification was a distinct possibility. The story goes that the academic offered to give the Alliance a positive writeup in a professional journal if it would—just once—decertify a farm. As we have reported, that has not been the Alliance's philosophy. Wille told us that in the past there had indeed been backsliding on some of Chiquita farms and that, after warnings, the farms got back in line. Chiquita confirms that both the certifiers and the company realize that the big changes take time; they are based on a model of continuous improvement. Nevertheless, all parties seem to be taking special pains to avoid even the appearance of a conflict of interest. They had better: the integrity of the relationship depends on it. Evolving Rainforest Alliance policy recognizes that there must be a firewall between the various services it offers.

As Rainforest Alliance executive director Tensie Whelan attested in May 2003:

> The Rainforest Alliance and associated members of the Sustainable Agriculture Network have been auditing Chiquita farms for more than ten years. We have seen remarkable progress toward more eco-friendly production and healthier working and living conditions. During 2002, we once again assessed all of the farms against our rigorous standards. Many of the audits were done without warning the farm managers in advance. I am pleased to report that all of the 117 farms were once again recertified . . .
>
> On average, we gave Chiquita farms lower scores this year as we continue to raise the bar for these farms that have long ago met the requirements of the core standards. Our program demands continuous improvement.
>
> Some Chiquita farms in Colombia and Costa Rica did not demonstrate the progress we expected, and farms in Honduras were decertified after a surprise audit uncovered two major nonconformances. These conditions were corrected and the farms were recertified before year's end.
>
> Chiquita and the company's farm managers have a record of compliance with our standards. We are now challenging them to dig deeper into persistent and systemic environmental issues such as water conservation, agrichemical use and biodiversity protection.

In the big picture, therefore, certification is only a tool, but a good one, in leveraging social and environmental reform. Starting from literally nowhere a scant fifteen years ago, timber and agricultural certification is now respectable in many circles. Even developing countries that were initially fearful that certification was simply another tactic of rich countries to rob them of market access are now cautiously considering its advantages.

Admittedly certification, even on the much larger scale envisioned by its advocates, will need substantial help from a range of institutions. The enormous problems of sustainable development, including the plight of the small farmer everywhere, and the dependency of many governments on the large exporting companies require the best efforts the world community can bring to bear. Some observers insist that food security hangs on the presence or absence of the big agricultural interests in these dependent countries. That a number of leading agricultural multinationals elect the Chiquita route, is, therefore, an outcome devoutly to be wished. Meanwhile, when a major agriculture company treats the environment kindly, provides jobs, and respects the needs and dignity of the workers, the larger questions of trade equity and land distribution become somewhat more tractable and less divisive. And isn't that what sustainable development is all about?

Some hold out hope that because Chiquita has taken these bold steps, its rivals in the United States, Latin America, and the Caribbean will sooner or later follow suit. But Noboa in Ecuador is a private company with little interest so far in going beyond the letter of the national labor or environmental law, and the markets have not punished that company. Nor has Dole lost market share, and the company continues to source millions of boxes of bananas from Ecuador. In mid-December 2002, billionaire David Murdock reached an agreement to purchase Dole, thus becoming the sole owner of the company. In accordance with the agreement, Murdock was to pay $1.4 billion for roughly 74 percent of the stock he did not already own. When combined with the company's debts of more than $1 billion, it puts the total price of the deal at about $2.4 billion. Deutsche Bank AG, Scotia Capital Bank, and Bank of America agreed to provide full financing for the deal. The move insulates Dole from any criticism or reform campaigns from investors. In fact, it means that two of Chiquita's main rivals and the two most important exporters from Ecuador can operate according to their own private whims. Among other consequences, the privatization takes the company completely out of reach of the social investment community.[8]

A New State of Play

After Chiquita made its move with the Rainforest Alliance, Dole and Del Monte quickly started certifying all of their farms to ISO 14001. In that sense, even if no other big company steps forward, Chiquita's risky decision in 1992 to certify with Better Banana has undoubtedly had the effect of raising the bar in a large segment of the banana industry. Also, as a result of those early decisions, leaders in the activist community are turning their energies to campaigns that reverse the unfortunate labor and environmental conditions on Ecuadorian farms.

Furthermore, Chiquita's pact with IUF and COLSIBA had the effect of strengthening the hand of the international labor movement. It is undeniably a significant advance when a major embattled company finally comes to terms with organized labor. Lagging companies are finding it much more difficult now to hide.

Meanwhile, Chiquita is moving the other way. As vice president, treasurer, and corporate responsibility officer, Jeff Zalla has a positive story to tell to the social investment community and, in fact, to all investors. "Some green or socially responsible investors have begun to show an interest in covering our company. For example, KLD Research & Analytics, Inc., does research for those firms and it has begun covering Chiquita." Perhaps more importantly, BB&T Capital Markets, a traditional equity research firm, visited the Costa Rican operations of Chiquita, Dole, and Del Monte, and in a report makes the link between the positive labor relations resulting from Chiquita's actions and worker productivity. As BB&T explains: "The farm worker is perhaps the most important element in the process, as only he/she is directly in contact with fruit and able to reduce waste and thereby increase productivity. The most effective means to reduce cost per box is to increase the yield per hectare."[9] BB&T found that over the past three years, simultaneously with its improved yield, Chiquita's average cost per box in Costa Rica declined more than one dollar—an extremely dramatic swing in the low-margin banana business. In March of 2003, BB&T upgraded its recommendation from buy to strong buy. But even casual day traders understand the value of productivity.

Zalla hopes investors recognize that corporate responsibility is only successful when it is bound into every operation. "At Chiquita, instead of having a large central corporate responsibility function, these issues are led by the top management group and overseen by a corporate responsibility steering committee. I chair the committee as corporate responsibility officer, but its members are from throughout the company—sourcing, logistics, marketing, human resources, environment, and public affairs, for example. We knew from the start that to be successful we needed leadership on values and standards in every area of the company. But we also knew that it has to start with the CEO. Fortunately Cyrus Freidheim is a very strong supporter. He sets the tone and expectations, and that helps drive the culture."

Zalla argues that achieving high social and environmental standards is also in the best interest of investors because it reduces risk and increases the certainty of future cash flows, which is ultimately the basis for any company's valuation. For example, achieving high labor standards can increase employee pride and productivity, but it also improves labor relations and reduces the risk of costly strikes. The result is beneficial to workers, the company, and investors. And it strikes Zalla as simply smart business, from both his perspectives—as corporate responsibility officer and as treasurer, in charge of risk management for the company.

Still, U.S. and European agricultural protectionism is a crucial part of the trade picture, and the arguments for change can be persuasive. In April 2002 Oxfam International launched a three-year campaign called Make Trade Fair, with the goal of changing the rules "so that trade can become part of the solution to poverty, not part of the problem." The Fair Trade label has a chance to benefit commercially from this kind of publicity. But Alistair Smith of Banana Link expressed his concerns in another e-mail: "Fair Trade needs to be careful to maintain its broad network of support in society, and not 'professionalize' to the point of becoming technocratic certifiers and little else. As usual, there are dangers and opportunities in rationalizing certification programs, but provided the minimum fair price and social standards are not diluted, then I see a positive future for Fair Trade. Danger is the big companies—especially retailers—trying to co-opt it and then dilute standards."

Chris Wille would probably agree. The major problem to address now, he says, is centralized buying. "Food chain power is consolidating at the supermarket. These buyers are influential and driving certification, but they are not demanding enough. They just want to cover their own butts. It's interesting that to some extent the buying decisions have moved ahead of the consumer. There is evidence in the coffee world to support this. Roasters are not convinced that consumers care, but are moving ahead anyway. With BB certification and Chiquita leading the way, we are in a new world of corporate responsibility. The chief information executive at Migros in Switzerland says the rainforests are back." In truth, unfortunately that is not the case, and complacency is probably the worst of all attitudes to assume toward these endangered ecosystems. As we outlined in the Introduction, considerable urgency is still warranted despite the remarkable progress by Chiquita and to a lesser extent its dollar rivals in managing the land sustainably where they operate. Deforestation and loss of species, soil, and water in the tropics continue.

While there has been a very promising beginning of labor peace on Chiquita farms, the plight of workers on Noboa and Dole farms in Ecuador continues to be deeply troubling. In mid-June 2003, seventy workers on the vast Los Alamos plantation owned by Noboa were summarily fired for the classic reason that they had formally presented a draft collective bargaining agreement to the Ecuadorian labor ministry a few days prior. The workers were all members and officials of unions formed recently. Union demands included social security registration (a legal requirement in any case) and a living wage. The presentation of the draft agreement was in full compliance with national law and with international labor conventions, whereas the firings were a clear violation of both. As we have noted elsewhere, this action by Noboa is but the most recent. Over the past year and a half the company has made a number of attempts to repress workers' efforts to organize. These workers and, in fact, most agricultural and textile laborers in developing countries are further bedeviled by protectionism in rich countries. Currently, the industrialized nations of the world spend about $1 billion a day on agricultural subsidies enacted at the behest of strong farm lobbies to keep cheaper exports from entering their countries.

Good News / Bad News

Meanwhile, corporate power, somewhat checked following the burst of the dot.com bubble and the rash of corporate scandals, is still a considerable factor in global politics. According to the fourth annual Corporate Social Responsibility Monitor survey of more than 21,000 consumers, shareholders, and corporate employees in twenty-one countries, significant proportions of people in most countries are unable to name a socially responsible company. People have a difficult time doing this despite consistently high expectations for companies to demonstrate such responsibility. Those surveyed also report strong interest in learning more about corporate social performance. The report describes high levels of consumer activism and empowerment around corporate social responsibility. The good news is that for the first time in four years of polling, consumers in developing countries are demanding more from companies in social and environmental areas. And in Japan, consumer attitudes and behaviors are finally moving toward higher North American and European levels. The bad news, however, is that in the United States recent corporate misdeeds and the resulting debates on corporate governance have lowered expectations for companies to go beyond the traditional economic role.

It must be said that purist capitalists have always insisted that the energy of corporate managers and workers should be dedicated to one goal alone: stock value. Any time spent on good works is a distraction. From a slightly different viewpoint, one could observe that the corporate responsibility movement is a tragic sign of societal decline. Have we reached such a low ethical point that corporations need to be rewarded for simply doing what we expect every citizen to do: behave honorably to our fellow humans and respect the ecosystems upon which life depends? In this sense, a corporate responsibility function in companies can be seen as a surprising but necessary corrective in a vast sea of business-as-usual and limited moral aspirations.

Unfortunately, throughout the world there are some questionable models of corporate influence and secrecy, peopled by some very powerful men and women. Perhaps at the top of the list is the Carlyle Group. The company is led by Frank Carlucci, former secretary of defense and deputy director of the CIA. Other high-powered names include George Herbert Walker Bush and James Baker, and, until they came to light after President George W. Bush declared a war on terrorism, several members of the bin Laden family. The company's own publicity describes the operation as "a vast, interlocking, global network of businesses and investment professionals" operating within the so-called iron triangle of industry, government, and the military. With $12 billion in funds under management, the Carlyle Group is well positioned to profit from all manner of high-level political policy decisions—including the war on terror.

Having well-connected board members is an ingrained corporate strategy and in itself is not illegal. (Carl Lindner's clout in the Clinton White House has already been documented in our narrative.) Chiquita's new board of directors has a number of executives with broad international experience and it also has Roderick M. Hills, former counsel to ex-President Gerald Ford and chairman of the Securities and Exchange Commission from 1975 to 1977, under President Ford. Hills is an expert on corporate governance. At various times in his career he has helped advance the Foreign Corrupt Practices Act (1977) and labored for heightened transparency of corporate accounting practices and for board of directors' independence. Hills's wife is Carla Hills, who served as U.S. trade representative under the first George Bush. In that position she was the chief U.S. negotiator for the North American Free Trade Agreement. NAFTA is notable for being the first trade agreement in which mechanisms were established to formally consider environmental and labor issues.

Individuals such as those we have mentioned no doubt would bring a great deal of confidence to any company with Chiquita's markets and holdings. However, with Chiquita's public commitment to core values and transparent reporting of progress, the company is clearly on the other side of a wide divide from an enterprise like Carlyle. It is perhaps safe to predict that Carlyle is unlikely any time soon to undergo a similar transformation.

Open Questions

Tensie Whelan says that the Rainforest Alliance plans to continue its entrepreneurial ways. Board and staff want to stay with the role they are playing as the way to get governments and institutions to work together. "We are small and nimble and not afraid to try new things," she says. "In our timber certification program we have fostered a new movement and have changed practice on the ground on more than 100 million acres. We have helped transform the banana industry. We have major plans for other agricultural products and tourism. We are unique among conservation organizations in our ability to get concrete results, acre by acre on the ground. Very few organizations get down to the level of detail we do. Now it's time for the customer and the retailer to step up to the plate."

And that is the main bottleneck at the moment. Jobbers, ripeners, supermarkets, and consumers need to be brought actively into the certification equation. Here and there this is starting to happen, but much more aggressive marketing is needed. It may take old-fashioned rabble-rousing from stakeholders who care about all three bottom lines, not just price and quality. The breakthrough might come if the Chiquita label were ever sufficiently transparent that shoppers could see beyond the famous female figure to the fruits of the extraordinary collaboration that lie behind it.

The timing and circumstances that made possible the events we have described here are undeniably unique. In that sense, ours is a once-told tale. But on the issue of transparency and commitment alone, Chiquita has brought an entirely new dimension to the concept of a responsible company. Together with the Rainforest Alliance, it has proved that certification can be a tremendously useful way for companies to differentiate themselves in the market, and that corporations can make a meaningful contribution to sustainable development. They have also proved that the road is not for the faint of heart. Few corporations are likely to follow precisely in Chiquita's path, but the way has been cleared for them to master their own fates. There is a real danger that some corporations will be surf riders, coasting along the same old track of risk-averse insularity, accounting sleight-of-hand, and CEO aggrandizement. In the worst cases they will add on a veneer of reform, becoming, sadly, the narrow, airless, priggish cultures of political correctness that have infected so much of daily life in the United States. We need to demand that companies rise above these deadening symptoms of our increasingly closed and inward-looking society. In a globalized market, Chiquita has proved that corporations can be imaginative, innovative, fair and transparent to stakeholders, and egalitarian in outlook. And that is not a bad description of the Rainforest Alliance, as well.

Unfortunately it has become painfully apparent that too many governments around that world—as presently constituted—cannot be trusted to tackle the challenges inherent in sustainable development. Transparency International, in its first Global Corruption Barometer survey developed with Gallup International in early July 2003, found that "if citizens had a magic wand the world over, they would most like to eliminate corruption from political parties." And that corruption hits the poor the hardest and the rich least.[10]

Far from revoking their corporate charters, we need companies to conceive, manufacture, grow, and market necessary goods and services that are useful to society. If they do, they will earn both respect and a justifiable profit. That is capitalism at its best. It is also an expression of true American roots. As the historian Henry Steele Commager put it: "America was born of revolt, flourished on dissent, and became great through experiment." It is time now for multinational companies (not least the huge retailing conglomerates) to learn to trust and to take some risk. A growing number of intensely interested people are watching closely.

Appendixes

Appendix A. Fundamental Conventions of the International Labour Organization

The International Labour Organization (ILO), established in 1910, promotes social justice and international standards on human rights and rights of workers. It has a tripartite structure for decision making that includes unions, governments, and employers' organizations that represent companies. The director general in 2004 is Juan Somavía. Labor standards are developed through conventions and recommendations. Governments that ratify the conventions are obligated to adhere to them. Eight conventions (of 180) are considered fundamental rights; these are summarized below.

No. 29—Forced Labor Convention (1930): Requires the suppression of forced or compulsory labor in all its forms.

No. 87—Freedom of Association and Protection of the Right to Organize Convention (1948): Establishes the right of all workers and employers to form and join organizations of their own choosing without prior authorization, and lays down a series of guarantees for the free functioning of organizations without interference by public authorities.

No. 98—Right to Organize and Collective Bargaining Convention (1949): Provides for protection against antiunion discrimination, for protection of workers' and employers' organizations against acts of interference by each other, and for measures to promote collective bargaining.

No. 100—Equal Remuneration Convention (1951): Calls for equal pay and benefits for men and women for work of equal value.

No. 105—Abolition of Forced Labor Convention (1957): Prohibits the use of any form of forced or compulsory labor as a means of political coercion or education, punishment for the expression of political or ideological views, workforce mobilization, labor discipline, punishment for participation in strikes, or discrimination.

No. 111—Discrimination (Employment and Occupation) Convention (1958): Calls for a national policy to eliminate discrimination in access to employment, training, and working conditions, on grounds of race, color, sex, religion, political opinion, national extraction, or social origin and to promote equality of opportunity and treatment.

No. 138—Minimum Age Convention (1973): Aims at the abolition of child labor, stipulating that the minimum age for admission to employment shall not be less than the age of completion of compulsory schooling.

No. 182—Worst Forms of Child Labour Convention (1999): Calls for immediate action against the worst forms of child labor.

Appendix B. Chemicals Banned under the Better Banana Program

Substances Subject to the PIC Procedure (UN Prior Informed Consent)
(By chemical CAS Number(s)—category—Pic Date)

2,4,5-T 93-76-5 Pesticide 1997

Aldrin 309-00-2 Pesticide 1991

Captafol 2425-06-1 Pesticide 1997

Chlordane 57-74-9 Pesticide 1992

Chlordimeform 6164-98-3 Pesticide 1992

Chlorobenzilate 510-15-6 Pesticide 1997

DDT 50-29-3 Pesticide 1991

Dieldrin 60-57-1 Pesticide 1991

Dinoseb/dinoseb salts 88-85-7 Pesticide 1991

EDB (1,2-dibromoethane) 106-93-4 Pesticide 1992

Fluoroacetamide 640-19-7 Pesticide 1991

HCH (mixed isomers) 608-73-1 Pesticide 1991

Heptachlor 76-44-8 Pesticide 1992

Hexachlorobenzene 118-44-8 Pesticide 1997

Lindane 58-89-9 Pesticide 1997

Mercury compounds (including inorganic mercury compounds, alkyl mercury compounds, and alkyloxyalkyl and aryl mercury compounds) Pesticide 1992

Methamidophos (soluble liquid formulations of the substance that exceed 600 g active ingredient/l) 10265-92-6 Severely Hazardous Pesticide Formulation 1997

Methyl-parathion [emulsifiable concentrates (EC) with 19.5%, 40%, 50%, 60% active ingredient and dusts containing 1.5%, 2%, and 3% active ingredient] 298-00-0 Severely Hazardous Pesticide Formulation 1997

Monocrotophos (soluble liquid formulations of the substance that exceed 600 g active ingredient/l) 6923-22-4 Severely Hazardous Pesticide Formulation 1997

Parathion [all formulations—aerosols, dustable powder (DP), emulsifiable concentrates (EC), granules (GR), and wettable powders (WP)— of this substance are included, except capsule suspensions (CS)] 56-38-2 Severely Hazardous Pesticide Formulation 1997

Pentachlorophenol 87-86-5 Pesticide 1997

Phosphamidon (soluble liquid formulations of the substance that exceed 1000 g active ingredient/1)13171-21-6 [mixture (E) and (Z) isomers) 23783-98-4 [(Z)- isomer] 297-99-4 [(E) isomer] Severely Hazardous Pesticide Formulation 1997

Appendix C. Selected Tables from Chiquita's 2001 Corporate Responsibility Report

Environmental performance *(owned banana divisions in Latin America)*
Summary of BBP Audits by the Sustainable Agriculture Network

Elements of the BBP Standards	1	2	3	4	5	6	7
Ecosystem conservation							
Natural habitats							
Reforestation		A				B	
Wildlife conservation							
Fair treatment & good working conditions							
Hiring practices							
Right to organize & freedom of association							
Occupational health		C		D		E	
Housing and basic services						F	G
Community relations							H
Integrated crop management							
Integrated pest management			I				
Permitted/prohibited agrichemicals						J	
Transport of agrichemicals							
Storage of agrichemicals	K					L	
Application of agrichemicals	M	N			O		
Integrated waste management							
Reduce							
Reuse							
Recycle						P	
Proper disposal						Q	
Clean appearance							
Conservation of water resources							
Protect rivers							
Rational use	R	R	R	R	R	R	R
Contamination	S	T					
Residual waters							
Monitoring							
Soil conservation							
Establishing new farms							
Erosion control			U				
Soil management							
Environmental planning & monitoring							

Appendix C

☐ In compliance
■ Non-compliance in actual performance (see footnotes)

1 – Santa Marta, Colombia   2 – Turbo, Colombia   3 – Costa Rica
4 – Guatemala   5 – Honduras   6 – Armuelles, Panama
7 – Bocas, Panama

A In Turbo, Colombia, a forested area between the production zones and a natural water source was less than the required five meters, the same as in the 2000 audit.

B In Armuelles, Panama, reforested areas between production zones and natural waterways remained too narrow.

C A worker in Turbo, Colombia, opened an insecticide-treated bag without having had a medical cholinesterase exam, an essential practice to ensure worker health. This was a violation of our policy and BBP requirements.

D In Guatemala, auditors noted that several packing stations had no system to prevent children from entering working areas.

E In Armuelles, Panama, several used cooking oil containers filled with drinking water were taken into the field. The use of inappropriate containers could result in accidental ingestion of a dangerous substance.

F In Armuelles, Panama, auditors found stagnant water and trash in a field near one of the community facilities, creating a mosquito breeding area and health hazard.

G In Bocas, Panama, families routinely had more people living in their homes than permitted by division policy. This overcrowding could lead to hygiene and waste problems.

H In Bocas, Panama, workers on neighboring farms sometimes walked through company farms and could have become exposed during agrichemical applications. Auditors found post-harvest organic waste, probably from the neighboring producers, on one of our farms.

I Costa Rica remains over-reliant on the use of agrichemicals for pest control. The division lacked a system to monitor and evaluate insect populations.

J In Armuelles, Panama, a shed had unused cases of paraquat, a "dirty dozen" pesticide. (We have not used paraquat since 1996. We returned the cases to the producer for disposal.)

K Agrichemical products without proper labels were found in a shed in Santa Marta, Colombia, along with soft drinks that workers were storing there improperly.

L In Armuelles, Panama, oil-based products were stored in a shed without proper ventilation.

M A worker in Santa Marta, Colombia, reached into an agrichemical container without proper safety equipment, leaving traces of the chemical on her sleeve. A worker was late posting warning signs in an area recently treated with agrichemicals and was not wearing the proper safety equipment.

N A worker in Turbo, Colombia, applied post-harvest fungicides without wearing all required safety equipment.

O Workers in Honduras entered an area sprayed with fungicides before the end of the required waiting period.

P Some farms in Armuelles, Panama, did not have a system to collect and recyle empty agrichemical containers. Workers cut up and buried empty containers in a compost trench.

Q A compost trench in Armuelles, Panama, had several non-conformances: a mixture of organic and inorganic waste, evidence of burned waste and no fence to prevent unauthorized access.

R All divisions lacked systems to reduce and recirculate water used in packing stations.

S In Santa Marta, Colombia, three fertilizer tanks did not have retention barriers to prevent overflow spills.

T In Turbo, Colombia, a company automotive maintenance facility was leaking oil, which could have entered a nearby stream.

U In Costa Rica, auditors noted a lack of progress in the establishment of groundcover plants to prevent soil erosion.

Social performance *(owned banana divisions in Latin America)*
Summary of internal assessments against Social Accountability 8000 Standard

Elements of SA8000 Labor Standard		1	2	3	4	5	6	7
Child labor								
Use of child labor	X				1	1		
Remediation policy	X							
Limits on young worker hours (age 15–17)	X				13			
Hazards to young workers	X				14			
Forced labor								
No forced labor	X							
Health and safety								
Accident and injury prevention	X	2	3		2	3	2	3
Senior manager accountable	X							
Regular and recorded training	X	4			4	4		
Systems to prevent and respond to threats	X							
Clean bathrooms, potable water, food storage	X				5	5	5	5
Housing	X				6	6		
Freedom of association								
Freedom of association	X				15			
Parallel means where law restricts	X							
No discrimination against union reps	X							
Discrimination								
Non-discrimination in employment	X				7		7	7
Respect practices and needs	X							
No sexually abusive behavior	X				16			
Disciplinary practices								
No mental or physical abuse or punishment	X							
Working hours								
Regular: max 48 hr. and 1 day off in 7	X				8			8
Overtime: max 12 hr., voluntary, paid extra	X	10			9	9		9
Compensation								
Meets legal min., ind. standards, basic needs	X				17		18	19
Convenient pay with full detail	X	11	12					
No use of labor contractors to avoid benefits	X							
Management systems	X							

■ In compliance
□ Absence of required policy or management system
■ Major non-compliance in actual performance (see footnotes)
☐ Not applicable
[X] Were unable to conduct SA8000 assessments for 2001 in Santa Marta, Colombia, due to safety concerns.

1 – Santa Marta, Colombia   2 – Turbo, Colombia   3 – Costa Rica
4 – Guatemala   5 – Honduras   6 – Armuelles, Panama
7 – Bocas, Panama

* Data from assessments conducted in February 2002

Multiple locations
1   In Guatemala and Honduras, contractors and suppliers were not screened for use of child labor.
2   In Turbo, Guatemala and Armuelles, several workers were not using required personal protection equipment when applying agrichemicals. In Turbo, chemical storage sheds did not segregate chemicals into different shelves. In Guatemala, inadequate precautions were taken to avoid spraying fungicides by airplane while workers were in the field.
3   In Costa Rica, Honduras and Bocas, some workers did not receive required medical exams. In Honduras and Bocas, some workers did not shower after applying agrichemicals. And in Honduras, children were loitering in a packing station.
4   In Turbo, Guatemala and Honduras, workers did not receive adequate safety training.
5   Guatemala and Armuelles did not monitor drinking water quality, and Honduras did not do so adequately. In both Guatemala and Bocas, one drinking water source had bacterial contamination. In Armuelles, inadequate facilities existed for the food storage.
6   Housing units were in poor condition in Guatemala and certain locations in Honduras, where bacteria also contaminated some drinking water tanks due to lack of maintenance.
7   Women lacked professional opportunities in Guatemala, Armuelles and Bocas. In Guatemala, temporary workers reported experiencing discrimination.
8   In Honduras and Bocas, workers did not always receive one day off in seven.
9   In Guatemala, Honduras and Bocas, hourly workers and administrators sometimes worked in excess of 60 hours. In Bocas, adminstrative personnel worked overtime with no additional compensation.

Turbo, Colombia
10 Ship loading personnel exceeded the maximum allowable workjing hours.
11 An independent contractor was two months behind in filing required social security documentation.

Costa Rica
12 Some workers did not understand their pay slips.

Guatemala
13 Some 16- and 17-year-old packing station workers exceeded the maximum allowable working hours, in violation of policy.
14 The division had not made a determination about the jobs safe for young workers.
15 Misclassification of some regular, full-time workers as temporary resulted in their not having the legal right to freedom of association.
16 Several female workers alleged sexual harassment.
17 Pay for temporary workers did not meet basic needs due to lack of benefits.

Armuelles, Panama
18 One independent contractor had not enrolled its employees in the national social security system.

Bocas, Panama
19 Some workers who arrived to work at the packing stations and were asked to come back in the afternoon were not paid for their initial time. (Management corrected this violation of local law immediately.)

Employee feedback on our Core Values *(owned banana divisions in Latin America)*
Farm workers

	2000							Elements of our Core Values		2001					
1	2	3	4	5	6	7			1	2	3	4	5	6	7

Integrity
Open communication
Honest communication
Straightforward communication
Lawful conduct
Living by our Core Values

Respect
Fair treatment
Respectful treatment
Recognize family
Individual differences
Cultural differences
Individual expression
Open dialogue
Sense of belonging

Opportunity
Personal development
Teamwork
Recognition

Responsibility
Pride in work
Pride in products
Pride in satisfying customers
In communities
In environment
Careful use of resources
Returns to shareholders

2000								2001						
50	63	64	42	20	33	18	*No. of workers surveyed*	0	50	60	89	71	42	39

☐ Very good   ▨ Good   ■ Bad   ■ Very bad

Ⅹ Unable to conduct assessments for 2001 in Santa Marta, Colombia, due to safety concerns.

1 Sant a Marta, Colombia   2 Turbo, Colombia   3 Cost a Rica
4 Guatemala   5 Honduras   6 Armuelles, Panama
7 Bocas, Panama

Appendix C

Employee feedback on our Core Values *(owned banana divisions in Latin America)*
Administrators and managers

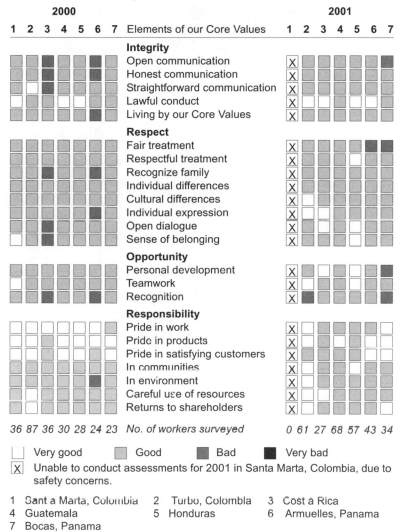

2000							Elements of our Core Values	2001						
1	2	3	4	5	6	7		1	2	3	4	5	6	7

Integrity
- Open communication
- Honest communication
- Straightforward communication
- Lawful conduct
- Living by our Core Values

Respect
- Fair treatment
- Respectful treatment
- Recognize family
- Individual differences
- Cultural differences
- Individual expression
- Open dialogue
- Sense of belonging

Opportunity
- Personal development
- Teamwork
- Recognition

Responsibility
- Pride in work
- Pride in products
- Pride in satisfying customers
- In communities
- In environment
- Careful use of resources
- Returns to shareholders

36 87 36 30 28 24 23 No. of workers surveyed   0 61 27 68 57 43 34

☐ Very good   ▨ Good   ▨ Bad   ■ Very bad

[X] Unable to conduct assessments for 2001 in Santa Marta, Colombia, due to safety concerns.

1 Santa Marta, Colombia   2 Turbo, Colombia   3 Costa Rica
4 Guatemala   5 Honduras   6 Armuelles, Panama
7 Bocas, Panama

Bananas sold to customers in selected European markets, 2001
(in thousands of boxes per week)

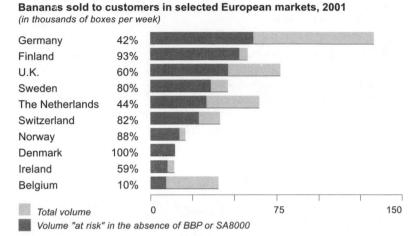

Germany	42%
Finland	93%
U.K.	60%
Sweden	80%
The Netherlands	44%
Switzerland	82%
Norway	88%
Denmark	100%
Ireland	59%
Belgium	10%

Total volume

Volume "at risk" in the absence of BBP or SA8000

Relative importance assigned by U.S. customers

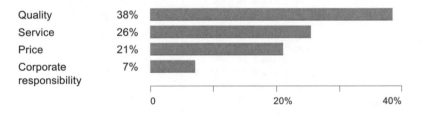

Quality	38%
Service	26%
Price	21%
Corporate responsibility	7%

Chiquita's Latin American banana operations — 2001

	Total employees	Owned volume as % of division exports	Total boxes shipped including purch. fruit (in millions)	Chiquita exports as % of country banana exports	Total payroll (in millions)
Santa Marta, Colombia	900	60%	4.4		$ 4.7
Turbo, Colombia	2,800	81	10.2	19%	15.5
Costa Rica	2,400	40	27.3	30	26.9
Guatemala	2,800	20	19.4	45	13.5
Honduras	3,500	75	11.8	46	16.2
Armuelles, Panama	3,200	85	6.5		15.9
Bocas, Panama	4,200	81	15.9	89	26.9
Subtotal	*19,800*	*55%*	*95.5*	*36%*	*$ 119.6*
Ecuador	30	0	8.8	4	0.4
Nicaragua	10	0	2.3	100	0.1
Total	*19,840*	*49%*	*106.6*	*23%*	*$ 120.1*

Appendix D. Sustainable Agriculture Network Members as of 2003

Conservación y Desarrollo, Equador

Fundación Interamericana de Investigación Tropical (Interamerican Foundation of Tropical Research, Guatemala

Fundación Natura, Colombia

Institute for Agricultural and Forestry Management and Certification (IMAFLORA), Brazil

Instituto Para La Cooperación y Autodesarrollo (ICADE), Honduras

Nepenthes, Denmark

ProNatura Chiapas A.C., Mexico

Rainforest Alliance, U.S. and Costa Rica

SalvaNATURA, El Salvador

Toledo Institute for Development and the Environment (TIDE), Belize

Notes

1. Risk, Transparency, and Trust

1. *Global Environmental Outlook 3,* produced by the UNEP GEO team; Division of Early Warning and Assessment (Nairobi, Kenya, United Nations Environment Programme, 2002).
2. Andrew Beattie and Paul R. Ehrlich, *Wild Solutions: How Biodiversity Is Money in the Bank* (New Haven: Yale University Press, 2001), p. 204. "The medical traditions of native forest peoples indicate that much remains for modern science to discover. The natives of the Amazon rainforest, for example, use no fewer than 1,300 plant species for curing various illnesses. Southeast Asian herbalists use some 6,500 species in their treatments. Some of these folk remedies may only be superstition, but if even a fraction of them have real medical value, they will be a boon to human health. Researchers suspect that as much as 10 percent of the estimated 90,000 plant species of Latin American forests will yield anti-cancer drugs. Recent tests in Costa Rica indicate that 225 of 1,500 plants tested there could be potential sources of anti-cancer drugs. But due to human activity, many of these plants may vanish before their value is ever known. Protection of these species depends upon the survival of forest habitat." (National Wildlife Federation website: www.nwf.org/keepthewildalive/periwinkle/bigpicture.cfm)
3. See Summary of the Thirty-Third Session of the International Tropical Timber Organization, November 4–9, 2002, *Earth Negotiations Bulletin,* vol. 24, no. 14 (International Institute for Sustainable Development, November 11, 2002), online at www.iisd.ca/linkages/forestry/itto/ittc33.

2. Red, White, and Bruised

1. Efficient refrigerator ships, called reefers, are a crucial variable in the banana business. Getting the fruit to market faster and in better shape than one's competitors can determine the amount of profit earned. A new top-quality reefer can cost $50 million.
2. The special section published on May 3, 1998, was a slap in the face for Carl Lindner personally, in that he was a onetime part owner of the paper and a Cincinnati icon.
3. The agriculture labels, initially ECO-OK, ran into trouble with Europeans, who allow the use of "ECO" for organic products only. The Alliance therefore changed the name of the project to Better Banana Program (BBP). It is now part of the Sustainable Agriculture Network (SAN), a consortium of Latin American nonprofit organizations and the Rainforest Alliance that is carrying out certifications throughout the region. The Alliance serves as the secretariat for SAN. Since 2002, all operations certified by the Alliance (timber, citrus, cacao, flowers, and tourist ships) say "Rainforest Alliance Certified" at the bottom of their labels to encourage brand recognition.
4. Schlesinger and Kinzer, *Bitter Fruit,* pp. 201ff.
5. McCann, *An American Company,* p. 233.
6. See the website www.totalresearch.com. Now allied with HarrisInteractive, for a number of years this company used interviews to rate brands according to a range of criteria. Chiquita placed number 9 out of more than three hundred brands, ahead of Levi Strauss at number 10 and below Reynolds Wrap foil at number 8. Kodak film claimed the top spot in the survey, which polled more than two thousand consumers, asking them to score brands on a scale of one to ten. Chiquita was the only food brand to make the top ten.
7. "Banana Growers Cited in Massive Fish Kill," *Tico Times,* July 17, 1992. At about the same time, the Costa Rican Forest Service accused Geest, another banana company, of illegally cutting down primary and secondary forests.
8. Nongovernmental organizations in the United States also receive government funding for carrying out projects under specific contracts, ostensibly not for political activities.

3. Times Change

1. Food and Agriculture Organization, Intergovernmental Group on Bananas and Tropical Fruits, Report of the Second Session.
2. Virginia Scott Jenkins has written in depth on the cultural aspects of the banana. *Bananas: An American History.* Quote on p. 159.
3. The English East India Company was founded in 1600; the Dutch East India Company, in 1602.
4. On a modern farm, the harvesting has become a somewhat gentler art. The stem is partially severed, then fairly gently placed on the workers' back before being cut completely off.
5. McCann, *An American Company,* p. 34. We have relied on McCann's colorful first-hand accounts for the perspective described in the following paragraphs.
6. McCann reports (pp. 39–40) that in 1952 the company owned 3 million acres of land, of which only 140,000 acres (about 5 percent) were actually planted in bananas. That year the company also had 102,000 acres in sugar crops.
7. Stanley, *For the Record.*
8. *The United Fruit Company in Latin America: Seventh Case Study in an NPA Series on United States Business Performance Abroad* (Washington, D.C.: National Planning Association, 1958). The May-Plaza book was published from research undertaken in 1955, a year after the CIA was involved in the overthrow of the Guatemalan government. American companies' performance in Latin America was the subject of much criticism at the time.

    The book includes a reproduction of a letter from José A. Mora, secretary general of the Organization of American States, to Charles J. Symington, head of the policy committee of the NPA, which was the sponsor of the work. In an interesting second paragraph Mora writes: "Your series of studies to date has clearly shown the United States enterprise is being moved more and more by a growing sense of social responsibility in its activities abroad and that it has demonstrated how profitable operation, on the one hand, and general economic improvement and development, on the other, can be mutually accelerative."
9. Dole had already introduced the variety. The change allowed the producers to abandon using copper sulfate, a principal cause of soil exhaustion that resulted in the abandonment of farms. Valery, more easily bruised, forced all of the companies to switch from shipping in bunches to shipping in boxes.

10. The Del Monte acquisition was in response to a takeover move by United Fruit. Fox, under pressure from Wall Street, had bought a block of stock in Del Monte. But Del Monte knew United Fruit was under a consent decree the company had signed with the U.S. Justice Department, whereby United Fruit was prohibited from increasing its share of the banana market. Del Monte went out and acquired the West Indies Fruit Company to avoid being taken over by United Fruit.

11. The company's prohibition preventing the ripeners from selling green bananas to any company not specified by United Fruit ultimately triggered a complaint to the EEC. An alienated ripener in Denmark claimed that the prohibition from selling across national boundaries was a restraint of trade within the new definitions established by the EEC. See Doc. 76/353/EEC: "Commission Decision of 17 December 1975 relating to a procedure under Article 86 of the EEC Treaty (IV/26699-Chiquita)." (Only the English text is authentic.) Official Journal L 095, 09/04/1976, pp. 0001–20.

12. Although the law is clear, the pay/benefits/contract situation in Ecuador is murky. Laws are commonly ignored, and workers resent having social security taken out of their pay, believing that it will not be available to them in the future. According to the Rainforest Alliance, workers compete for jobs at the Reybancorp farms, where the working conditions are better now; there is safety, good equipment, showers, bathrooms, clean lunchrooms with nutritious food, medical and dental care. These benefits, the Alliance believes, take the sting out of the low hourly rate called for by the government. Reybancorp Agricolas S.A. is the banana-producing subsidiary of the Holding Favorita Fruit Company. Its farms produce 26 percent of the total exported by Reybanpac. The fruit is exported under the Favorita brand name.

13. Traci Carl, "Era Ends in Latin America as Banana Industry Slides: Protracted Trade War Hurt Chiquita's Ability to Compete," *Washington Post*, December 2, 2001.

4. Why Bananas and Why Chiquita?

1. The European Union has also strongly criticized the United States for legislation passed in May 2002 that provided large farm subsidies primarily benefiting industrial agriculture, and started on a path to limit subsidies under its Common Agriculture Policy. The EU and the U.S. are both being chastised by the developing nations: all of this is a major topic on the agenda of the ongoing WTO Doha trade round.

2. See Jenkins, *Bananas,* pp. 20–21.
3. Emilio Fernandez, "Fugitive Banker Back in Mexico after Extradition," Reuters, August 2001.
4. Langley and Schoonover, *The Banana Men,* p. 35.
5. Radio contact solved a thorny problem in the precision movement of bananas to market. See also www.ketupa.net/nbcII.htm.
6. Schlesinger and Kinzer tell the Guatemala story superbly in their 1982 exposé, *Bitter Fruit,* based in part on secret documents pried from the U.S. government under the Freedom of Information Act.
7. Bernays, a nephew of Sigmund Freud, was widely recognized as a public relations genius.
8. A principal in one of the company's public relations firms at the time remembers that Robert Kennedy, then his brother's attorney general, pressured United Fruit to let the U.S. government use company vessels to carry pharmaceutical supplies to Cuba. When the company turned him down, Kennedy threatened to inspect the company's books, and that finally got some action.

5. Strange Bedfellows

1. Newsom, "Achieving Legitimacy?"
2. For the conference papers, go to www.bananalink.org.uk.
3. The Pesticide Action Network in April 2002 called for the worldwide phasing out of Paraquat. It is banned or severely restricted in seven European and four developing countries. The antibanana group Foro Emaús and the National University raised the alarm on Paraquat in Costa Rica in the early 1990s. In a 2001 publication, Foro Emaús mentioned that Chiquita was the only banana company not using Paraquat. Bravo, a chlorinated isophthalic acid derivative, is registered for use as a fungicide to control a wide variety of fungal diseases in crops.
4. Tucker, *Insatiable Appetite,* p. 172.
5. In conversations with Alliance certifiers, we subsequently learned that they were not happy about the planting of ficus because it is not a local species.
6. We noted at the time that the recycled wooden pallets went to Europe; a program for pooling used pallets in the United States was being explored. How sensitive even the private companies are to market response is demonstrated by Noboa's experience in Ecuador years ago. When the company tried to use white packing cases for Germany instead of the usual light brown, the Germans immediately objected

because of the chlorine content in the white. Noboa, without hesitation and despite the cost, returned to the natural brown.

7. Geest was at one time a major British player in the banana business, exporting from former colonies in the Caribbean. In 1992 it acquired virgin land in the Sarapiquí region of northeast Costa Rica and established eleven plantations that required clearing of primary forest. In 1995 Geest's entire banana division was sold to a fifty-fifty joint venture of Fyffes and the new Windward Island company WIBDECO. The Costa Rican plantations were quickly resold to a Central American consortium, currently trading as Caribana.

8. Mauna Kea, Platenara Rio Sixaola and the EARTH model banana farm in Costa Rica (see note 9) were the first farms to achieve certification.

9. EARTH (in English, the Agricultural College of the Tropical Humid Region) is a major campus-like agricultural research and teaching facility originally funded by the U.S. Agency for International Development and the Kellogg Foundation. The school studies innovative ways to use alternative, renewable resources and waste materials. EARTH was a pioneer in the development of alternative papers from banana stalks, and it has carried out pilot projects for recycling plastics and building homes from a bamboo-like grass called Caña Brava. On its grounds, EARTH maintains a prototype banana farm where experimentation takes place. Its holdings include a rainforest reserve with trails.

6. Grass-Roots Snapshots

1. During the long period when the principal banana producers were cultivating the Gros Michel variety, "Panama disease" was a serious problem and copper sulfate was the fungicide of choice.

2. World Bank, "Adjustment Lending and Mobilization of Private and Public Resources for Growth" (Washington, D.C.: World Bank, 1992). As wages declined and poverty spread during the 1980s, the Costa Rican government was cutting social service expenditures in line with World Bank and IMF requirements to reduce the current account deficit. A 1992 World Bank report on adjustment lending shows government budgets for health, nutrition, and sanitation programs declining 35 percent in real terms between 1979 and 1988, while real expenditures per capita fell by 45 percent during the same period. The government has had to postpone new investments, such as the construction of clinics and schools, while the demand for social services continues to rise with the increased impoverishment of the population.

3. The term "ombudsman" originated in Scandinavia. It originally meant something similar to an ambassador: a person or office established to safeguard the rights of individual citizens (or a particular group of citizens) in relation to the powers and actions of government. In Sweden an ombudsman for justice was appointed in 1809. In the more than forty countries that now have general ombudsman offices, their most common functions are to receive and investigate complaints, to act as a spokesperson, and to advocate improvements for individuals or specific groups.

4. The current president of CI is Russell A. Mittermeier, a distinguished biological anthropologist. CI has projects in a number of Latin American countries. Its mission is "to conserve the earth's living natural heritage, our global biodiversity, and to demonstrate that human societies are able to live harmoniously with nature."

5. Listed below are some of the environmental and natural-resource laws in Costa Rica as reported in a profile of the country published by LEAD, a Rockefeller Foundation leadership project:

Water Law No. 276 (1942): Refers to the use of the territorial seas, lagoons and rivers, groundwater, and water from pluvial origin. Potable water is under Health Ministry jurisdiction, and a special potable water law prohibits any construction, facility, or activity near a potable water source. Water law also provides protection for forests near water resources.

Penal Code: Relates to the environment when dealing with usurpation, environmental damage, and use of illegal substances or explosives for fishing, law violations, and authority disobedience.

National Parks Service Law No. 6084 (1977): Provides regulations on activities in the national parks.

Wildlife Conservation Law No. 7317 (1992): Provides protection and use regulations for wildlife, declares public domain of the wild fauna, and declares for public interest the wildlife, conservation, research and development of genetic resources, species, and wild varieties. It also specifies the legal obligations of the General Wildlife Office of MIRENEM, prohibits fishing and hunting of endangered species, and creates a Wildlife National Register Office. This law also prohibits the dumping of polluting agents into any water source, and provides rules for wildlife reproduction and trade.

Forestry Law No. 7174 (1990): Provides protection for forestry resources; declares protected zones for all forests and forestry lands, especially areas near water sources; provides rules for forest exploitation; and punishes illegal tree cutting and intentional forest fires.

National Health Law No. 5395 (1973): Provides protection against pollution, solid, toxic and hazardous wastes.

Mining Code: Gives total state domain in relation to mineral resources. Provides rules for mineral exploitation and exploitation concessions; prohibits mineral exploitation in national parks and biological reserves.

Coastal Zones Law No. 6043 (1977): Provides regulations for use, exploitation, tourism, and other activities in the coastal zone. Provides protection to coastal resources.

Decree 7210: Declares all mangroves and other salt-water forests as forest reserves.

Decree 20265: Prohibits national and international wild flora trade, especially orchids, ferns, bromeliads, and cactus, exempting those legally produced by artificial reproduction systems.

7. Blue Bananas

1. Two years later, the FAO Codex Alimentarius adopted a standard (adding also qualifications regarding heavy metals and hygiene) that expanded the definition of quality to include size and color.
2. EC Control Regulations for BOCEC Nr. 2898/95.
3. In Europe, governments promoting projects to assist communities in developing countries often fund nonprofit organizations, including labor groups. European ties to former colonies drive these projects as much as a more highly developed sense of commitment to poor countries than exists so far in the United States government.

8. Agricultural Antagonists

1. Mort Rosenblum, *Goose in Toulouse* (New York: Hyperion, 2000), p. 73.
2. He has since become an itinerant militant, and Agence France press reported that he was blocked from going to Qatar for the November 2001 meeting of the WTO until the French government intervened. He subsequently served a brief prison term for the McDonald's mischief, and he continues to get widespread notice in the press as he mows down fields of genetically modified crops and visits one or another courtroom. In November 2002, Bové was sentenced to six months in jail for his attack on genetically modified rice and another eight months for an earlier protest (Reuters, November 21, 2002).
3. Observers have said that the EU banana regime pitted the United States

against the European Union; banana-crazy Germany against France, Spain, and Britain; U.S. multinational giant Chiquita against two powerful European rivals; U.S. investment interests against U.S. diplomatic niceties; Guatemala against Costa Rica; Latin American growers against Caribbean growers, and the U.S. government against four major Latin American nations.

4. Center for Global Food Issues, Churchville, Va., 1999.

5. Letter to the Editor, *Philanthropy* magazine, December 21, 1998. Reprinted by permission of the Philanthropy Roundtable, www.philanthropyroundtable.org.

6. From an unpublished paper prepared in 2001 for the Center for Rural Affairs in Walthill, Neb.

7. Report of a meeting May 23, 2002, on "High Tech vs. Low Tech Farming for Feeding the World." For the full report, see www.ngocongo.org/ngo subs/sustdev.docs.htm.

8. Melissa Jacobs, an organic farmer and traditional seed saver who operates a 60-acre organic farm in New York State, said during the NGO meeting at the UN: "Less than 2 percent of the population in North America is growing the food that we all eat . . . In New York State alone we are losing three thousand farms per year." Robert Goodman, the University of Wisconsin professor, mentioned a study sponsored by the Rockefeller Foundation. It found that in 2000, 90 percent of the food produced in the world was consumed locally and only 10 percent was internationally traded, a figure that had not changed since 1960.

9. An article in the *New Scientist* in September 2002 reported that the World Food Program had been delivering genetically modified food as emergency aid for the last seven years, in spite of the fact that African countries concerned about GM-free exports to Europe have been rejecting food that might be genetically modified. Their fear is that grain fed to livestock grown for export to Europe as organic beef might be "contaminated."

10. "Last Days of the Banana," *New Scientist*, January 18, 2003, pp. 27–28.

11. One needs to be careful in thinking about fungicide applications to control Black Sigatoka. Often any increase is due mostly to the higher use of protectant, rather than to systemic fungicides. The protectant used by Chiquita is a Class IV (practically nontoxic) agrichemical. The company insists it continues to monitor latest developments in plant science to identify possible breakthroughs that might allow it to achieve adequate control of Black Sigatoka with fewer agrichemicals.

12. See www.fas.usda.gov/pecad/highlights/2002/01/Brazil; also Lak-samana.net.

9. "Daylight Come . . ."

1. "First Time Report from Chiquita Impresses on Every Count," *Business and the Environment,* vol. 12, no. 11 (November 2001), p. 5. The editor went on to say that the report "hurtles Chiquita out of obscurity into the highest echelon of corporate reporters."
2. "Human Rights Watch Report Reveals Child Labor and Workers' Rights Abuses in Ecuador's Banana Fields," IUF press release, April 25, 2002, posted on web site www.iuf.org.
3. Tucker, *Insatiable Appetite,* ch. 3.
4. Bucheli, "United Fruit Company in Colombia," ch. 5, p. 20.
5. According to its 2000 corporate responsibility report, Chiquita had 85 percent of its workers represented by Permanent Committees (p. 69).
6. In fact, the government *has* enacted laws and regulations to protect unions and union leaders. They also "prevent Permanent Committees from being used as 'antiunion devices.'" Further, in labor bargaining, the Ministry of Labor recognizes a union over a Permanent Committee. Solidarity Associations are not legally allowed to represent workers in collective bargaining (2000 corporate responsibility report, p. 70).
7. Discrimination by gender is forbidden under ILO convention no. 111. The ILO conventions are central to the SAI social certification program as well (see Chapter 11 and Appendix A).
8. In Panama the social accountability assessors for SA 8000 reported that employees were not allowed to take breast-feeding breaks. The division's action plan includes an initiative to discontinue this prohibition.
9. At that time, on a typical banana plantation in Honduras, Chiquita had roughly 1,500 plants per hectare, of which about one third might have the blue chlorpyrifos-saturated bags enclosing the bunches. Those bags have a cumulative total of 5.1 grams of chlorpyrifos per hectare. This dosage is more than a thousand times less than the maximum dosage allowed by the EPA in a food handling establishment.

10. The Many Faces of Corporate Responsibility

1. For an interesting discussion of this point, see Esty, *Greening the GATT: Trade, Environment,* and *the Future.*

2. These standards boosted the level of quality assurance throughout the electronics industry. In those days any customer could assess its suppliers. The waste inherent in this process was apparent. It was broadly believed that the appraisals should be weighed against common standards of quality assurance. This climate opened the door to third-party inspection and to the establishment of private commercial assessment organizations. Like BBP, ISO 14001 can be a de facto enforcer of national laws since the standards always reference them. But under ISO 14001, unlike BBP, only the auditor and the company know that the management system honors the laws.

   These first standards for quality assurance were seen as contractually binding. During the 1970s the debate among major industries in the United States, Europe, and Japan moved to how best to inspect and assure that the standards were being met. The U.K. government, with the intention of having British quality standards gain respect in world markets, took its homegrown quality standard (BS 5750) to the international standards community. The nongovernmental national standards bodies are coordinated through the International Organization for Standardization (ISO) in Geneva. Its role is to promote the development of standardization and related activities to facilitate the international exchange of goods and services.

3. Malcolm Baldrige was secretary of commerce from 1981 until his death in a rodeo accident in July 1987. His help in drafting one of the early versions of a congressional quality improvement act caused the award created by the measure to be named after him. The Baldrige Award is given by the president of the United States to businesses that are judged to be outstanding in a number of areas. It was envisioned as a standard of excellence that would help U.S. organizations achieve "world-class" quality.

4. ISO 14001 is the proper terminology when referring to the environmental management system standard. ISO 14000 is used only when referring to all the standards produced by the ISO Technical Committee on Standards; in other words, the ISO 14000 series of standards. Today virtually everyone sticks with ISO 14001.

5. Since agricultural companies did not participate significantly in the formulation of ISO 4001, many felt that farms were an awkward fit for a process that was basically conceived for electronics and other manufacturing industries. Adjustments were made to accommodate the differences, and Rudy Amador, environmental affairs regional manager Latin America for Dole, wrote in an e-mail that his firm was the first agricul-

tural company in the world to certify to the standard, beginning with the certificate in June 1998 by Dole's farm, Standard de Costa Rica. After that, certified environmental management systems were implemented in all the Latin American countries where Dole has fresh fruit operations. It operations have also been certified in the Philippines, Thailand, and Spain. And interest appears to be growing in the agricultural sector. (In 2000, we met with Dole's general manager in Costa Rica, Peter Gilmore, along with Amador, the company's Latin American certification coordinator, and Dole legal coordinator Juan Carlos Rojas Zeledon. Later, Amador amplified some of the Dole policy and history by e-mail.)

6. The formulators of the standard, including representatives from organized labor, recognized that in these industries, with crews going out to platforms, for example, for weeks at a time, working hours and other labor issues could not be monitored. Agriculture was another area put aside initially, with the knowledge that seasonal workers would be a problem for a global set of labor standards. Instead, in consultation with ILO and others, it was agreed to treat agriculture as a special case with its own working group and its own guidance document for certifiers who focus more on the "spirit" than the "letter" of adherence to ILO conventions. It was obvious to the formulators that auditing to the letter of the law would mean that most farms could never be certified. The compromise obviously was not universally satisfactory.

7. Certified companies, called SA 8000 signatories, are mostly apparel and toy companies; many are located in Asia. Although SA 8000 is still relatively new, as of 2002 the organization said it had certified 133 sites in 27 industries in 26 countries. See www.sa-intl.org. On October 7, 2003, Social Accountability International recognized Chiquita's achievements and partnership with the Rainforest Alliance by honoring them with its Corporate Conscience award. The competitive award honors outstanding social and environmental performance and seeks to encourage responsible business practices.

8. The San Francisco nongovernmental organization had been a crucial resource for Chiquita at each stage in the development of its corporate responsibility program.

9. While we could not cover the operations of the "Great White Fleet," we were told the crews of these ships, usually from the Philippines, often remain aboard for as long as forty days with no shore leave, though they receive extended leave afterward. Like oil rig crews, these workers cannot be monitored and thus have gained an exemption under SA 8000. Chiquita opted for ISO 14001 for its shipping operations.

11. Leveling the Label Field

1. Fair Trade is gaining a toehold in North America as well. Its coffee is being promoted in California, Canada, and elsewhere in coffee shops supported by young latte drinkers.

2. The former chancellor of West Germany, Willy Brandt, headed an independent commission that was the first to raise this concern in a public document.

3. Some reviewers of this book, which was written by Eduard Douwes Dekker under the pen name Multatuli, found it a "savagely funny" satire and one of the great Dutch literary classics. It became a bible for the colonial reform movement and the plight of small coffee farmers in Java and elsewhere. Ten years after its publication, Captain Lorenzo Dow Baker brought his first shipload of bananas from Jamaica to the United States.

4. The Fair Trade Labeling Organization International (FLO) coordinates and maintains the registry of certified producers. The basic criteria for a Fair Trade label are as follows:
   Direct trading links with producers in developing countries;
   Guaranteed minimum price to producers (country specific);
   Price premium to producers;
   Credit allowances or advance payments; and
   Long-term relationships.

5. "Best of the Bunch," pamphlet published by Banana Link, U.K., n.d.

6. FAO Committee on Commodity Problems, Intergovernmental Group on Bananas and on Tropical Fruits, Report of the Second Session.

7. FAO Committee on Commodity Problems, Intergovernmental Group on Bananas and the Market for "Organic" and "Fair-Trade," Report of the First Session.

8. www.wn.elib.com/Steiner.

9. In Switzerland in 1930 a politician, Helmut Müller, gave impetus to this new movement. His goals were at once economic, social, and political, envisioning a much more direct link between the production and consumption stages. In the 1950s, organic farming established itself in France, driven by doctors and others concerned with food and its effect on health. Two groups were prominent: a movement linked to commercial operators supplying inputs to producers (Lemaire-Boucher) and an independent, noncommercial NGO (the Nature and Progress Association).

10. The IFOAM Accreditation Programme is the practical realization of the

federation's commitment to harmonize an international guarantee of organic integrity. IFOAM established the International Organic Accreditation Service Inc., an independent organization, to run its accreditation program. The Programme, including the IFOAM Seal (see chapter 11 opening illustration), is held by IFOAM.

12. Unfinished Business

1. The data are from a banana industry newsletter, *Sopisco News* (Panama: Nova Media Publishing Co., no. 19, 2002).
2. This information, Smith explained, comes from the Caribbean Media Corporation in Bridgetown. The figures for 2002 exports are incomplete and thus do not add up, but this is what was available as of May 2003: St. Vincent—36,900 tons (1997–2001 average was 37,500); St. Lucia—44,700 for first nine months, estimating 60,000 for year (1997–2001 average was 59,860); Grenada—500 for first nine months, estimating 700 for year (1997–2001 average was 420 tons); Dominica—estimating 17,400 tons for year (1997–2001 average was 28,100).
3. SASA members, a subset of the larger iSEAL group, include SAI (Social Accountability International), SAN, the Fair Trade Labeling Organization (FLO), and the International Federation of Organic Agriculture Movements (IFOAM). A steering committee met for the first time in mid-2002 in New York at the invitation of Business for Social Responsibility. It was BSR's agricultural working group that organized the meeting. (Chiquita's Jeff Zalla was recently elected to the board of BSR.) Unilever, Starbucks, Green Mountain, and Nestlé were among those in the audience as the four main labelers set forth their organizational stories and their hopes for SASA. Both Dole and Chiquita are involved in the larger stakeholder meetings.
4. The accreditation process is tied to EN 45011 and its international equivalent, ISO 65 (The International Organization for Standardization Guide 65: General Requirements for Bodies Operating Product Certification Systems). The standards contain provisions relating to the structure of inspection/certification bodies. As a consequence, and in order to achieve a level playing field, all imports of produce from outside the European Community for sale as organic must also be certified by bodies complying with either or both of the standards.
5. After early criticism by some growers that retailers dominated the scheme, the organization added members from throughout the supply chain. It now characterizes itself as a supply-chain partnership.

6. Another of the acronyms, FROG, stands for First Raise Our Growth. It represents a scenario in which activists insist that standards set at global summits be enforced. Developing countries resist, arguing that if the developed nations demand high environmental standards, they should "First Raise Our Growth!" Some nations leapfrog from under-developed status to first mover, especially in areas of technology. Bjorn Stigson said, "People in western nations respond in uneven ways—sometimes by offering help in improving the environment, and sometimes in raising various cries of 'FROG!' themselves." See "Exploring Sustainable Development: WBCSD Global Scenarios 2000–2050," summary brochure (Geneva, 1997).

7. Anheir, Glasius, et al., *Global Civil Society, 2001:* also at www.ise.ac.uk/depts/global/yearbook. The number of national groups in France, for example, had soared by 1996 to fifty-four thousand; in Bangladesh there were ten thousand. Weiss and Gordenker, *NGOs, the UN, and Global Governance,* p. 7.

8. Chiquita's other main rival, Del Monte, not known for its social or environmental progressiveness, might be distracted at the moment. A December 2002 lawsuit brought by a group of Mexican investors demanded that the sale six years ago of Del Monte be rescinded. The suit alleges that bribery and corruption were involved in the sale of the banana giant at a "ridiculously low price."

9. "Fresh Fruit Producers: Field Trip to Costa Rica Yields New Insights," August 26, 2002. Report to investors.

10. Transparency International press release, July 3, 2003.

Selected Bibliography

Anheir, H., M. Glasius, et al., eds., *Global Civil Society, 2001.* Oxford: Oxford University Press, 2001.

Avery, Dennis T. *Saving the Planet with Pesticides and Plastic: The Environmental Triumph of High-Yield Farming.* Indianapolis: Hudson Institute, 2000.

Bucheli, Marcelo. "United Fruit Company in Colombia: Labor Conflicts, Political Relations, and Local Elite, 1899–1970." Ph.D. diss., Stanford University, 2001.

——*United Fruit Company in Colombia, 1899–2000.* New York: New York University Press, forthcoming.

Cashore, Benjamin, Graeme Auld, and Deanna Newsom. *Politics Within Markets: Regulating Forestry Through Non-State Environmental Governance.* New Haven: Yale University Press, in press.

Esty, Daniel C. *Greening the GATT: Trade, Environment, and the Future.* Washington, D.C.: Institute for International Economics, 1994.

Food and Agriculture Organization, Committee on Commodity Problems, Intergovernmental Group on Bananas and the Market for "Organic" and "Fair-Trade." Report of the First Session, Gold Coast, Australia, May 4–8, 1999.

—— Intergovernmental Group on Bananas and on Tropical Fruits. Report of the Second Session, San José, Costa Rica, December 4–8, 2001.

——The Market for "Organic" and "Fair-Trade" Bananas. FAO doc. CCP: BA/TF 01/5.

Global Environmental Outlook 3: Past, Present, and Future Perspectives. Nairobi: UNEP, 2002.

Global Exchange. Bibliography compiled by Steve Striffler, University of North Carolina, www.globalexchange.org/economy/bananas/bibliography.

Grossman, Lawrence S. *The Political Ecology of Bananas: Contract Farming, Peasants, and Agrarian Change in the Eastern Caribbean.* Chapel Hill: University of North Carolina Press, 1998.

Holliday, Charles O. Jr., Stephan Schmidheiny, and Philip Watts. *Walking the Talk: The Business Case for Sustainable Development.* London: Greenleaf Publishing, 2002.

Identity and Struggle at the Margins of the Nation-State: The Laboring Peoples of Central America and the Hispanic Caribbean. Aviva Chomsky and Aldo Lauria-Santiago, eds. Durham, N.C.: Duke University Press, 1998.

Human Rights Watch. "Tainted Harvest." New York, 2002. www.hrw.org.

Jenkins, Virginia Scott. *Bananas: An American History.* Washington, D.C.: Smithsonian Institution Press, 2000.

Kell, Georg, and John Ruggie. "Global Markets and Social Legitimacy: The Case of the 'Global Compact.'" Paper presented at an international conference at York University, Toronto, Ontario, Canada, 1999.

Korten, David C. *When Corporations Rule the World.* New York and San Francisco: Kumarian Press and Berrett-Koehler Publishers, 1995.

Langley, Lester D., and Thomas Schoonover. *The Banana Men: American Mercenaries and Entrepreneurs in Central America, 1880–1930.* Lexington: University Press of Kentucky, 1995.

Leipziger, Deborah. *SA 8000: The Definitive Guide to the New Social Contract.* London: Pearson Education, 2001.

Litvin, Daniel. *Empires of Profit: Commerce, Conquest, and Corporate Responsibility.* New York: Texere, 2003.

Márquez, Gabriel García. *One Hundred Years of Solitude.* New York: Harper and Row, 1970.

May, Stacy, and Galo Plaza. *The United Fruit Company in Latin America: Seventh Case Study in an NPA Series on United States Business Performance Abroad.* Washington, D.C.: National Planning Association, 1958.

McCann, Thomas P. *An American Company: The Tragedy of United Fruit.* New York: Crown Publishers, 1976.

Moeberg, Mark, and Steve Striffler, eds. *Bananas, Conflict, and Capitalism in Latin America and the Caribbean.* Durham, N.C.: Duke University Press, forthcoming.

Moran, Theodore H. *Beyond Sweatshops: Foreign Direct Investment and Globalization in Developing Countries.* Washington, D.C.: Brookings Institution, 2002.

Newsom, Deanna. "Achieving Legitimacy? Exploring Competing Forest Certification Programs' Efforts to Gain Forest Management Support in the U.S. Southeast, Germany, and British Columbia, Canada." M.S. thesis, Auburn (Alabama) University, n.d.

Schlesinger, Stephen, and Stephen Kinzer. *Bitter Fruit: The Story of the American Coup in Guatemala,* expanded ed. Cambridge, Mass.: Harvard University, David Rockefeller Center for Latin American Studies, 1999.

Stanley, Diane. *For the Record: The United Fruit Company's Sixty-Six Years in Guatemala.* Guatemala: Centro Impresor Piedra Santa, 1994.

Striffler, Steve. *In the Shadow of State and Capital: The United Fruit Company, Popular Struggle, and Agrarian Restructuring in Ecuador, 1900–1995.* Durham, N.C.: Duke University Press, 2002.

Tucker, Richard P. *Insatiable Appetite: The United States and the Ecological Degradation of the Tropical World.* Berkeley: University of California Press, 2000.

Vandermeer, John, and Ivette Perfecto. *Breakfast of Biodiversity: The Truth About Rain Forest Destruction.* Oakland, Calif.: Institute for Food and Development Policy, 1995.

Vernon, Raymond. *In the Hurricane's Eye: The Troubled Prospects of Multinational Enterprises.* Cambridge, Mass.: Harvard University Press, 1998.

Weiss, Thomas G., and Leon Gordenker, eds., *NGOs, the UN, and Global Governance.* Boulder, Colo.: Lynne Rienner, 1996.

Index